HYDROCLIMATE
The Influence of Water Movement on Freshwater Ecology

HYDROCLIMATE
The Influence of Water Movement on Freshwater Ecology

IAN R. SMITH

Edinburgh, Scotland, UK

ELSEVIER APPLIED SCIENCE
LONDON and NEW YORK

ELSEVIER SCIENCE PUBLISHERS LTD
Crown House, Linton Road, Barking, Essex IG11 8JU, England

Sole distributor in the USA and Canada
ELSEVIER SCIENCE PUBLISHING CO., INC.
655 Avenue of the Americas, New York, NY 10010, USA

WITH 17 TABLES AND 135 ILLUSTRATIONS

© 1992 ELSEVIER SCIENCE PUBLISHERS LTD

British Library Cataloguing in Publication Data

Smith, Ian R.
Hydroclimate: the influence of water movement on freshwater ecology.
I. Title
574.52632

ISBN 1-85166-724-5

Library of Congress Cataloging-in-Publication Data

Smith, Ian R.
Hydroclimate: the influence of water movement on freshwater
ecology/Ian R. Smith.
    p.   cm.
Includes bibliographical reference and index.
ISBN 1-85166-724-5
1. Freshwater ecology. 2. Hydrometeorology. I. Title.
QH541.5.F7S65 1992   91-31565
574.5′2632—dc20   CIP

Preface

Sun, wind and water draining from the land interact with the morphological features of a water body to create the environment experienced by freshwater plants and animals. The result of this interaction can be considered as the freshwater *hydroclimate* and this plays the same role as that of conventional climate in terrestrial ecology. Agriculture, for example, has long been supported by specialist meteorological services which not only provide farmers with a sound interpretation of weather and climate without excessive technicality but which also consider relations between climate and the growth of crops and stock. There is a need for a similar service in freshwater ecology and applied biology. This book is the result of a number of years devoted to developing part of that service. It concentrates on the influence of all forms of water movement on the ecology of fresh waters.

Water movement implies interest in both the quantity of water moving through river basins which reflects the climate of the catchment as well as the nature of the fluid motion within the rivers and lakes of the basin. The book is not so much a review of recent research as an attempt to establish a logic—how knowledge of water movement can contribute to understanding the ecology of fresh waters. Two points follow directly. One is the need for the book to be comprehensible to non-specialists and I hope this is achieved by emphasising simple physical arguments despite the rather formidable list of symbols. There is also the difficulty that, once you start discussing freshwater ecology, when do you stop? The chapter on freshwater ecosystems is, inevitably, over-simplified and the intention is simply to provide a basis for assessing the influence of physical conditions. The general theme, however, is that of stimulating ecologists to consider the implications of the environmental features described than to suggest that these are fully understood already. I am very conscious that there is a gap between this account of some of the physical features of rivers and lakes and the interests and activities of practising ecologists. I hope this book will help to bridge that gap.

I hope the book will appeal to all with an interest in rivers and lakes—undergraduates, graduate students and practising professionals in both management and research. I also like to think that the book will be of interest to naturalists. Understanding processes is always satisfying and, to me, the motion of eddies in a river is just as delightful as the flight of a dipper. Engineers and hydrologists may find the account of the freshwater environment presents their own interests in a new light and give them an insight into the ecological consequences of their activities.

I would like to thank Peter Maitland, Chris Gibson and Alan Steel. All read parts of the book and made a number of helpful suggestions. Above all, I would like to thank Alec Lyle who read the whole draft, supplied most of the photographs and who, over the years, made working with rivers and lakes such a pleasure. The responsibility for the errors is, obviously, mine. I am also grateful to Ken Morris and David Jones for allowing me to use their photographs.

IAN R. SMITH
18 Tarpin Road, Edinburgh, EH13 0HW, UK

Contents

Preface v

Chapter 1 **Hydroclimate** 1

Chapter 2 **The structure and dynamics of river basins** 4
2.1 River basin structure 4
   2.1.1 Introduction 4
   2.1.2 Longitudinal river sections 6
   2.1.3 Catchment form 8
   2.1.4 River networks 12
2.2 Climate and streamflow 16
   2.2.1 Runoff generation 16
   2.2.2 Climate classification 19
   2.2.3 Average flow 22
   2.2.4 Seasonal variation in flow 25
   2.2.5 Discharge frequency 26
   2.2.6 Floods and low flows 31

Chapter 3 **Outline of fluid dynamics** 39
3.1 Basic features of fluid motion 40
   3.1.1 Characteristics of fluid motion 40
   3.1.2 Methods of analysis 42
3.2 The nature of turbulence 43
   3.2.1 Basic features 43
   3.2.2 Boundary layer flow 48
   3.2.3 Turbulent structure 51
   3.2.4 Advection–diffusion equation 54
   3.2.5 Free turbulence 56
3.3 Hydrodynamic properties of particulate matter 58
   3.3.1 Introduction 58
   3.3.2 Static properties 60

3.3.3   Bed deposits   62
3.3.4   Particles in suspension   62
3.4   The dynamics of particulate matter   64
3.4.1   Introduction   64
3.4.2   The onset of bed movement   66
3.4.3   Particles in suspension   67
3.4.4   Shields' diagram   70
3.4.5   Cohesive sediments   72

Chapter 4   **The dynamics of hydraulic systems**   74
4.1   Linear systems theory   74
4.1.1   Introduction   74
4.1.2   Response to standard inputs   77
4.1.3   Mass balance of a mixed basin   79
4.2   Incomplete mixing   82
4.2.1   Mixing processes   82
4.2.2   Residence time distributions   83
4.2.3   Reaction efficiency   87
4.3   Compartment models   88
4.3.1   Introduction   88
4.3.2   Parallel systems   89
4.3.3   Series systems   91
4.3.4   Systems with backflow   93
4.3.5   Seasonal stratification   95
4.4   Final remarks   97

Chapter 5   **Hydraulic characteristics of rivers**   98
5.1   River morphology   98
5.1.1   Introduction   98
5.1.2   Classification of river morphology   98
5.1.3   Relations between channel dimensions and flow   . .   101
5.1.4   Particle size variation along a river   108
5.1.5   Micro-topography of river beds   111
5.2   River mechanics   113
5.2.1   River as a boundary layer flow   113
5.2.2   Stage discharge curves   123
5.2.3   Dissolved and suspended loads   124
5.2.4   Numerical example   126
5.3   Hydrodynamic features of river habitats   131
5.3.1   Introduction   131

5.3.2 River stability 132
5.3.3 Micro-hydraulics 136
5.3.4 Dispersion and mixing in rivers 139

Chapter 6 **Hydraulic characteristics of lakes** 142
6.1 Lake morphology. 142
  6.1.1 Introduction 142
  6.1.2 Standard measures 142
  6.1.3 Hydrodynamic measures 146
6.2 Wind characteristics. 148
  6.2.1 Introduction 148
  6.2.2 Features of land station data 150
  6.2.3 Wind over water 151
6.3 The hydraulic structure of lakes 155
6.4 Wind–water interactions. 157
  6.4.1 Basic processes 157
  6.4.2 Empirical observations. 158
  6.4.3 Langmuir circulations 159
  6.4.4 Hydrostatic balance 161
  6.4.5 Vertical current structure 162
6.5 Motion in isothermal lakes. 163
  6.5.1 Introduction 163
  6.5.2 Two dimensional steady circulation 165
  6.5.3 Circulation in irregularly shaped lakes 167
  6.5.4 Surface seiches 169
6.6 Motion in stratified lakes 173
  6.6.1 The occurrence of stratification 173
  6.6.2 Properties of a stratified profile 176
  6.6.3 Thermocline tilt and internal seiches 178
  6.6.4 Regimes of stratified motion 179
  6.6.5 Steady state circulation in stratified lakes 183
6.7 Surface waves 183
  6.7.1 Wave characteristics. 183
  6.7.2 Deep water waves. 186
  6.7.3 Shore zone waves 189
6.8 Other features of lake motion 191
  6.8.1 Inflow and outflow dynamics 191
  6.8.2 Turbulence in lakes 195
  6.8.3 Exchanges between sediment and water 199
6.9 Final remarks 202

Chapter 7   **River basin modification** 204
7.1   Reservoirs and flow regulation 204
   7.1.1   Types of reservoir 204
   7.1.2   Storage theory 206
   7.1.3   Reservoir features 208
7.2   Modifications to river habitats 209
   7.2.1   Introduction 209
   7.2.2   Effects of flow changes 210
   7.2.3   Changes in river structure 211

Chapter 8   **Freshwater Ecosystems** 215
8.1   Idealised ecosystems 215
   8.1.1   Ecosystems and ecosystem models 215
   8.1.2   Idealised river ecosystems 219
   8.1.3   Idealised lake ecosystems 222
8.2   Rate controlled growth 224
   8.2.1   Introduction 224
   8.2.2   Resources 224
   8.2.3   Regulators 226
   8.2.4   Physical losses 228
   8.2.5   General remarks 229
8.3   Event controlled growth 229
   8.3.1   The nature of events 230
   8.3.2   Species response to events 231
   8.3.3   Hydrodynamic stress and population loss 234
   8.3.4   Population loss due to a sudden discharge of toxin . 235
8.4   Space controlled growth 237
   8.4.1   Spatial variation in plankton growth 237
   8.4.2   Habitat availability 239
8.5   Final remarks 240

Chapter 9   **Synthesis** 242
9.1   Background 242
   9.1.1   Interests involved in river basins 242
   9.1.2   The logic of hydroclimate investigations 244
   9.1.3   Modelling aquatic systems 246
9.2   River models 246
   9.2.1   River description 247
   9.2.2   Transport and storage 249
   9.2.3   River ecology 253

9.3 Lake models 256
   9.3.1 Introduction 256
   9.3.2 Basic features of lake hydroclimate 256
   9.3.3 Lake ecology 259

References 261

Appendix 1: **The properties of water** 271
  A.1 Viscosity–temperature relationship 271
  A.2 Density–temperature relationship 271

Appendix 2: **Notation** 273

Index 279

1

Hydroclimate

If we consider hydroclimate to be concerned with the influence of all forms of water movement on the ecology of fresh waters, then it is reasonable to ask what is the substance of hydroclimate, what is the justification for the listed contents. There are, clearly, two strands involved—understanding the nature of water movement in rivers and lakes and finding ways of incorporating that understanding into ecological investigations. Except in mountain headwaters, all freshwater environments are inter-related, all are affected to some extent by what happens upstream. To understand the influence of water movement, therefore, it is necessary to consider both transport and storage within a river basin as well as how hydrodynamic processes influence the characteristics of particular sites.

These arguments are the justification for the listed contents which are, essentially, repeated in Fig. 1.1 but where the intention is to stress the logical structure. The first step is to outline the basic features of a river basin. This is, potentially, a major project in itself but it is important to highlight the main features of catchment structure, climate and streamflow which do determine, ultimately, the nature of the aquatic environment.

The nature of river and lake habitats is considered in Chapters 5 and 6 but, before doing so, the technical background, the basic science of water movement, is discussed. Fluid dynamics considers the essential features of turbulent motion that can be applied in both rivers and lakes as well as those properties of particulate matter relevant to its behaviour in turbulent flow. The chapter on the dynamics of hydraulic systems is concerned with the bulk movement of water rather than its hydrodynamic properties. It provides the basis for understanding transport and storage in river basins. The movement of water through

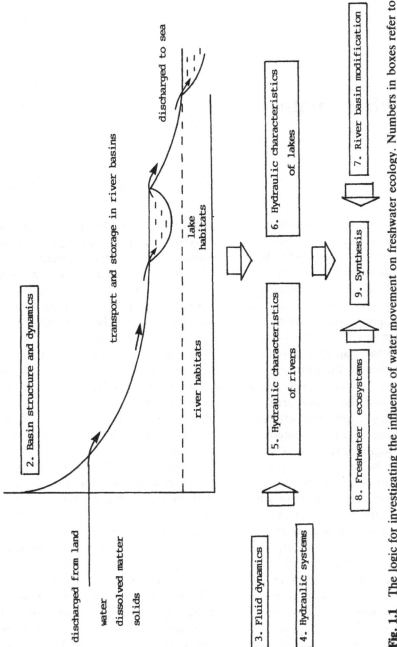

Fig. 1.1 The logic for investigating the influence of water movement on freshwater ecology. Numbers in boxes refer to chapter numbers.

river basins is somewhat similar to the flow of electricity through simple circuits and some elementary linear systems theory is included. Chapters 5 and 6 themselves require little justification. They form the core account of how flowing water interacts with morphological features to create freshwater habitats.

The justification for Chapter 7 is the recognition that few river basins are in a pristine state. Quite apart from pollution, natural flows are stored, regulated and diverted and the natural river and lake morphology altered by engineering works. The emphasis is on how these factors alter the physical conditions rather than on reviewing the extensive literature on the impact of development.

The chapter on freshwater ecosystems attempts to establish a logical basis for assessing the influence of physical conditions. No attempt is made to summarise the content of freshwater ecology other than to outline idealised accounts of river and lake ecosystems which form the background to understanding the various ways growth rates can be affected.

The last chapter links the various strands together. How the influence of physical conditions lies within the conflicting interests involved in river basin management is considered as well as the rationale of physical investigations in support of ecology. The bulk of the chapter, however, is concerned with technical synthesis, particularly the possible bases for mathematical modelling both to predict hydroclimate and to link physics and ecology.

Little is said about instrumentation and the specific features of hydroclimate observations. The variety of circumstances is so great that generalisations are of doubtful value. The physical situation, the nature and level of detail of the investigation and the resources available—both finance and technical skill—can all have an influence. The idea of a standard 'underwater weather station' is attractive but may be impractical.

Finally, it is worth emphasising what is not considered. Radiation and water temperature are discussed only in so far as they affect motion in stratified lakes and water chemistry only appears in relation to the affects of dilution and mixing. Nor are any major environmental issues, such as climate change, discussed directly. Obviously, if changes in streamflow, wind and temperature are postulated, then the potential change in the freshwater environment can be followed through.

2

The Structure and Dynamics of River Basins

A particular freshwater habitat cannot be considered in isolation. Because of the movement of water through the system, there is always some dependence on what happens upstream or in the contributing catchment area. There are no absolute boundaries and the only natural unit is the entire river basin. It is appropriate, therefore, to start by examining some of the features of river basins—their topography and the movement of water through them. Not only do these determine the nature of freshwater environment but they also provide the link between fresh waters and the general geography of a region.

2.1 RIVER BASIN STRUCTURE

2.1.1 Introduction

Many accounts of freshwater studies start with a map of the catchment and a longitudinal section of the main river (Fig. 2.1). Besides indicating the location, scale and general features of the study area, such diagrams provide much useful information at a rather general level. The map demonstrates the pattern of water flow within the catchment, the location of lakes and major tributaries and the extent of the contributing area at any point. The longitudinal section indicates barriers to fish movement, river slope is a major determinant of habitat characteristics and the profile may also give some indication of climate. Rainfall is often closely correlated with altitude and the average air temperature falls 6°C for every 1000 m increase in altitude.

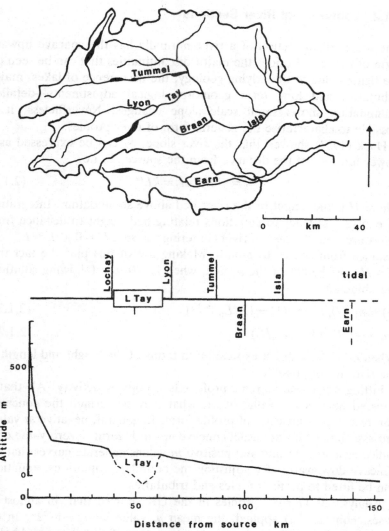

Fig. 2.1   The catchment area and longitudinal profile of the River Tay (based on Maitland and Smith, 1987).

Topographical maps emphasise the particular features of a river basin and obscure the underlying pattern. Some quantitative analysis of catchment topography is justified since some ecologically useful measures of catchment characteristics do emerge.

2.1.2 Longitudinal River Sections

The longitudinal section of a river normally has the concave upward form of Fig. 2.1. Besides the major discontinuities that can be seen on the figure—due to underlying geology, the occurrence of lakes, major tributaries and long term geomorphological adjustments—detailed examination shows small scale slope changes. Nevertheless, it is possible to characterise the general form of river profiles.

Hack (1957) showed that the river slope, S, can be expressed as a power function of the distance from the source, L, i.e.

$$S = -dH/dL = kL^m \tag{2.1.1}$$

where H is the height of the river bed above some datum. Integration of eqn (2.1.1) leads to equations relating bed height to distance from the source. In the case of rivers entering the sea, $H = 0$ at $L = L_m$, the distance from source to mouth. Making use of this plus the fact that $H = H_s$, the height of the source, when $L = 0$, the following solutions are obtained

$$-1 < m < 0, H = H_s[1 - (L/L_m)^{m+1}] \tag{2.1.2a}$$

$$m = -1, H = k \ln(L_m/L) \tag{2.1.2b}$$

When $m < -1$, a direct expression in terms of the height and length of the river is not possible.

Fitting equations to river profiles is a popular activity. All that is stressed here is the value of m, what may be termed the concavity parameter, as a measure of profile form. In general, negative m values are associated with the usual concave upwards form, a zero value of m indicates a straight line and positive m values generate curves that are concave downwards. By adjusting the boundary conditions, equations can be fitted to partial profiles and tributaries.

Analysis of 17 river profiles in the UK showed that, in almost all cases, eqn (2.1.2a) applied, the mean m value being -0.725. In the case of the River Tay, separate equations were fitted above and below Loch Tay, the m value above the Loch being -1.01 and below -0.23. The River Danube, which is approximately 2800 km long, has a value of m equal to -1.58. Figure 2.2 compares observed and computed profiles in the River Tweed.

If eqn (2.1.2) provides a reasonable description of the river profile, then not only does the m value give a measure of profile concavity but it can be used to generate what is, effectively, the frequency

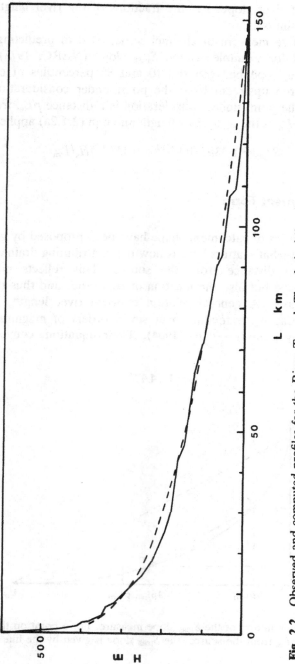

Fig. 2.2 Observed and computed profiles for the River Tweed. The dashed line is the fitted curve relating the height above sea level, H, to the distance from the source, L, i.e., $H = 564(1 - (L/154)^{0.27})$.

distribution of river slope, i.e. the fractions of the river length with various slope limits.

An alternative measure of channel slope, used in predicting flood characteristics for example, is the S_{1085} slope (NERC, 1975). It is defined as the slope between the 10 and 85 percentiles of channel length measured upstream from the point under consideration (see Fig. 2.3). If the point under consideration is a distance pL_m from the source where L_m is the total river length and eqn (2.1.2a) applies, then

$$S_{1085} = 1 \cdot 33 p^m (0 \cdot 9^{m+1} - 0 \cdot 15^{m+1}) H_s / L_m \qquad (2.1.3)$$

2.1.3 Catchment Form

Various measures of catchment shape have been proposed by geomorphologists but what matters here is how the contributing drainage area increases with distance from the source. This reflects not only catchment shape but also the location of tributaries and thus the flow within the basin. A general relation between river length, L, and catchment area, A_c, extending over seven orders of magnitude was established by Leopold *et al.* (1964). Their equation, converted to metric units, is

$$L = 1 \cdot 54 A_c^{0 \cdot 6} \qquad (2.1.4)$$

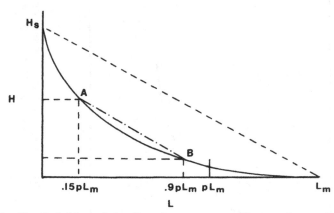

Fig. 2.3 The definition of the S_{1085} slope measure. For a point on the river a distance pL_m from the source, the S_{1085} slope is given by the line AB.

This equation is based on a comparison between rivers and simply indicates that length increases with catchment area but not in proportion. The fact that the exponent is 0·6 and not 0·5 indicates that catchments become more elongated with increasing size.

The converse relation within a catchment can be expressed as a Gomperz curve, i.e.

$$A_c = a e^{-be^{-cL}} \qquad (2.1.5)$$

Analysis of a number of British rivers indicates that all three parameters are related to the overall catchment dimensions, A_m and L_m, the drainage area and length at the mouth, i.e.

$$a = 0·98 A_m + 510 \qquad (r = 0·957) \qquad (2.1.6)$$

$$b = 0·07 A_m/L_m + 6·23 \qquad (r = 0·847) \qquad (2.1.7)$$

$$c = 4·50/L_m \qquad (r = 0·985) \qquad (2.1.8)$$

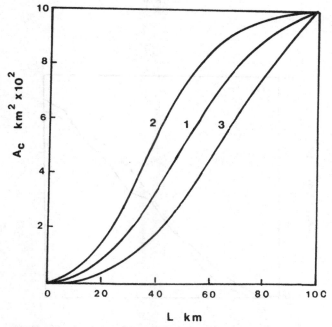

Fig. 2.4 The effect of the parameter c on the relation between catchment area, A_c, and the river length, L. For a river of length 100 km, the computed value of c, using eqn (2.1.8), is 0·0441 (curve 1). Curve 2 is generated with the value of c increased to 0·06 and, in curve 3, the value of c is reduced to 0·035.

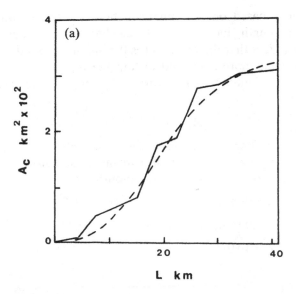

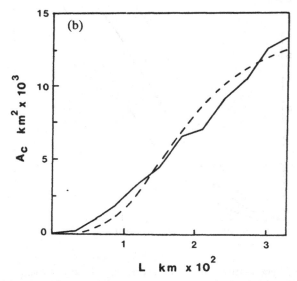

Fig. 2.5 Observed and computed catchment area, A_c, in terms of distance from the source, L. The full lines indicate measured catchment areas and the dashed lines the fitted curves. (a) River Tyne (Lothian, Scotland), $A_c = 356e^{-6.61e^{-0.108L}}$; (b) River Thames (England), $A_c = 13\,614e^{-8.69e^{-0.014L}}$.

Parameter a is, essentially, a scaling factor ensuring that $A_c = A_m$ at the mouth and parameter b does not vary greatly. It is parameter c that exerts the greatest influence on the length–area relation. A value of c greater than that calculated from the regression equation increases curvature while a lower value reduces it (see Fig. 2.4). Figure 2.5 compares observed and computed relations.

A knowledge of the altitudinal distribution within the contributing catchment is valuable since altitude partly determines land use and climate. Conventionally, this information is presented as dimensionless hypsometric curves (see Fig. 2.6). Strahler (1954) gives a large number

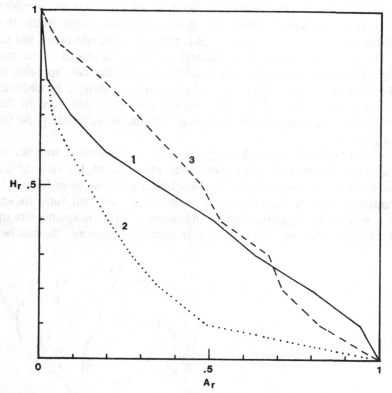

Fig. 2.6 Dimensionless hypsometric curves at points on a river. The relative height, H_r, is given by $(H' - H)/(H_p - H)$ where H' is the height within the catchment, H is the height of the river bed at a distance, L, along the river and H_p is the highest point in the catchment. The relative area, A_r, is equal to A'/A_c where A' is the area of the catchment where the height is greater than H' and A_c is the catchment area at distance, L, from the source.

of examples as well as a rather unsatisfactory curve fitting procedure. The value of such curves is that, in non-dimensional form, different topographies have characteristic shapes. Curve 1 on Fig. 2.6 is typical of a large number of basins while the others represent the likely extremes. Curve 2 is characteristic of a saucer-shaped basin, i.e. flat land around a lake and enclosed by a ring of hills, while the almost straight line form of curve 3 represents dissected plateau topography.

2.1.4 River Networks

The plan view of a river network has a more orderly structure than a casual glance at a map suggests. The simplest analysis of river networks is based on Horton's procedure as modified by Strahler (1952). A stream without tributaries is defined as first order, below the confluence of two first order streams is defined as second order and so on. The entry of a first order tributary into what is already a second order stream has no effect on the numbering, a similar rule applying at higher orders (see Fig. 2.7). A number of more elaborate procedures have since been derived but the original scheme is that used in the river continuum concept (see Section 8.1.2) and is quite adequate for ecological use.

The fundamental network property is that, if the total number of streams of different order in a river basin are counted, the ratio of the number of streams of order u, to the number of those of order $u + 1$, is constant. This is the bifurcation ratio, β. A constant ratio means that a plot of the logarithms of the numbers of streams against stream order is a straight line (see Fig. 2.8). In general, therefore, the number

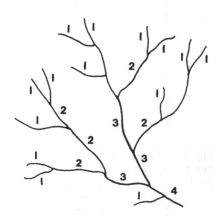

Fig. 2.7 Stream order number system.

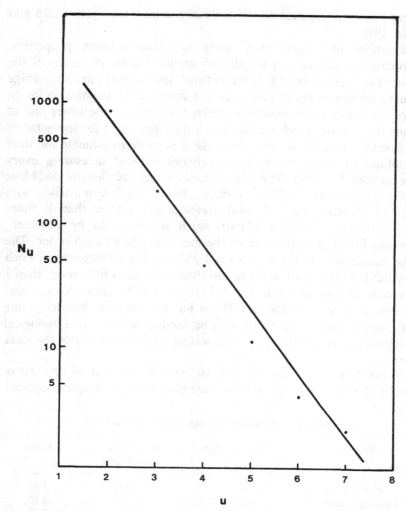

Fig. 2.8 The relation between the stream order, u, and the number of streams of order u, N_u, within the River Tay Basin, based on an analysis of 1:50 000 maps. The best fit relation is $N_u = 7692e^{-1.197u}$, corresponding to a bifurcation ratio of 3·31.

of streams of order u, N_u, is given by

$$N_u = ae^{-bu} \quad (2.1.9)$$

The parameter b is equal to $\ln \beta$, the bifurcation ratio, which usually has a value of between 3 and 5. The actual numbers and the highest order observed depends on the scale of the map used (see Smith and Lyle, 1979).

Relations of similar form apply to other network properties, particularly mean stream length and drainage area. A survey of the River Tay basin showed that the network laws also applied to average stream width and depth (Table 2.1). Failure of the length relation to apply at higher order numbers reflects the fact that the lower end of many rivers is drowned, i.e. there is insufficient space for the network to develop fully. These relations make it possible to estimate the total quantities of water in the river network without measuring every stream (see Fig. 2.9). The overall basin totals are: length, 5827 km; water surface area, $34 \cdot 5$ km^2; volume, $20 \cdot 7$ m^3 $\times 10^6$. Comparison with Fig. 2.1 is interesting. The total stream length is more than 40 times the main river length (137 km) which is, as might be expected, virtually identical to the sum of the mean lengths of each order. The total surface area of the network is $1 \cdot 35$ times the surface area of Loch Tay but the volume of water in the river network is little more than 1 per cent of that of Loch Tay ($1597 \cdot 5$ m^3 $\times 10^6$). These figures are, obviously, specific to the Tay Basin but the analysis, besides giving estimates of basin-wide totals, can be used as a basis for a biological sampling procedure that gives due weight to the structure of the river basin.

Network analysis has obvious defects. Estimation of the mean length of a first order stream, for example, involves lumping together

Table 2.1 Properties of the River Tay network

Property	Relationship	Correlation coefficient	Ratio[c]
Stream numbers	$7\,692e^{-1 \cdot 197u}$	0·990	3·31
Drainage area	$0 \cdot 23e^{1 \cdot 30u}$	0·871	3·67
Stream length[a]	$0 \cdot 26e^{0 \cdot 92u}$	0·988	2·51
Stream width[b]	$0 \cdot 51e^{0 \cdot 72u}$	0·992	2·05
Stream depth[b]	$0 \cdot 097e^{0 \cdot 33u}$	0·961	1·39

[a] Order 1–5 only. For orders 5, 6 and 7, lengths are virtually the same (approx. 30 km).
[b] Width and depth are based on data collected during a survey in April when the flow was below average.
[c] The ratio of the value appropriate to a particular order to that in the adjacent order. The number of streams decreases with order but all other values increase.

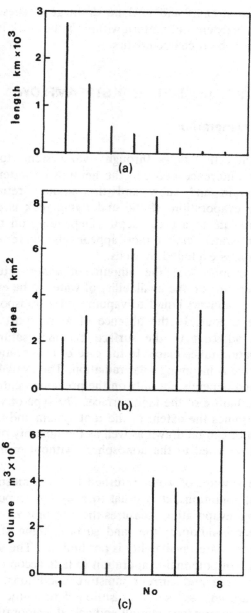

Fig. 2.9 Properties of the River Tay Basin network. (a) Total stream length in each order; (b) total water surface area in each order; (c) total water volume in each order.

mountain source streams and lowland drainage ditches. Even more importantly, the existence of pattern within the network itself does not imply pattern in habitat characteristics.

2.2 CLIMATE AND STREAMFLOW

2.2.1 Runoff Generation

Runoff, the water that flows through river systems to the sea, is, essentially, the difference between the amount of water falling on a catchment as rain and snow and the amount returned to the atmosphere by evaporation. If the underlying rock is porous, some water may percolate to a great depth. Depending on the geological structure, such water may either appear elsewhere or be stored underground unless exploited by wells.

The actual evaporation, the amount of water returned to the atmosphere, depends on the availability of water to be evaporated, an energy supply to convert liquid to vapour and a transport mechanism to remove the vapour. In the absence of a transport mechanism— wind—the air adjacent to the surface becomes saturated and no further evaporation takes place. In the case of evaporation from land, the energy source is incoming solar radiation. The availability of water to be evaporated depends not only on the previous weather conditions but also on the nature of the land surface. The type of vegetation, for example, determines the extent of the root system and thus the depth from which water can be drawn as well as the quantity of water stored on leaves and returned to the atmosphere without ever reaching the soil.

Because the amount of water returned to the atmosphere depends partly on local conditions, it is usual to consider evaporation in two stages. Potential evaporation measures the evaporative capacity of the current weather conditions, the land surface being standardised as short grass where water availability is not limiting. The second stage is the modification of potential evaporation to take account of the nature of the land surface and current moisture conditions. The latter is normally measured as soil moisture deficit—the accumulated difference between previous rainfall and actual evaporation.

In the case of evaporation from a water surface, the question of water availability, obviously, does not arise. With large water bodies, however, the energy source for conversion from liquid to vapour can be altered. When the lake temperature is rising, more radiation goes into heating the water and evaporation is suppressed but, during the cooling phase, some of the heat released is used as an energy source for evaporation. The annual cycle of evaporation from a lake may be markedly out of phase with the incoming radiation.

The process of runoff generation, particularly how it varies with time, can be visualised by considering a section through the ground surface (Fig. 2.10). What happens to rainwater depends not only on

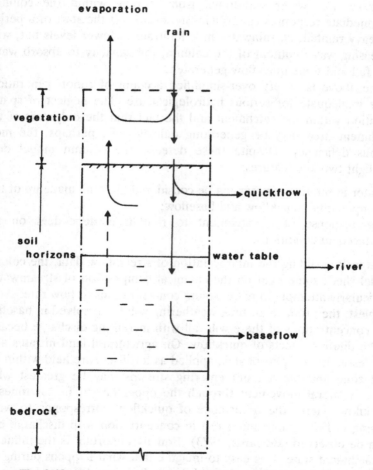

Fig. 2.10 Diagrammatic section through a catchment surface.

the structure of the column—the possible pathways through it—but also on its state. At the start of a period of rainfall, the state of the column depends on the amount and distribution of water within it but itself changes as the rain continues to fall.

If the rain is sufficiently intense and the initial water content high, then the rate at which water arrives at the surface is greater than the capacity of the lower soil horizons to receive it and the excess flows through the upper soil layers direct to the stream system. This is referred to as quickflow. At the other extreme when no rain has fallen for some time, the only source of supply for the stream system is baseflow, i.e. water withdrawn from storage within the column. Intermediate responses can, obviously, occur. At the start of a period of heavy rainfall, all rainwater may infiltrate to lower levels but, with increasing water content of the column, the capacity to absorb water may fall and some quickflow generated.

The above is a very over-simplified account of runoff generation, quite inadequate for serious hydrological use. The neglect of spatial variation within the catchment and the fact that the only part of the catchment area may be generating quickflow is, perhaps, the most serious deficiency. Despite these defects, the column model does highlight two key features:

—water in the river system can be considered as being made up of two components—quickflow and baseflow;
—the response of a catchment to rainfall is dependent on the antecedent conditions.

As well as aiding the interpretation of streamflow data, the column model sheds some light on the chemical composition of streamwater, particularly attempts to relate solute concentrations to flow rate. At its simplest, the products of rock weathering will be dissolved in baseflow and concentrations of these will fall with increasing discharge because of the diluting effects of quickflow. On agricultural land nitrates and, to a lesser extent, phosphates, applied as fertilisers are held within the root zone and the amount entering streams may be greatest when there is lateral movement through the upper layers, i.e. at times of quickflow. Here, the occurrence of quickflow corresponds to water flowing in field drains and a rise in concentration with discharge may then be observed (Edwards, 1973). Equally important is the influence of catchment state. It is easy to imagine that what happens during the first major flood after a prolonged dry spell is very different from what happens after weeks of continuous rain. The influence of catchment state on water chemistry is discussed in more detail by Walling and Foster (1975).

The resulting interaction between weather and catchment characteristics is seen in a hydrograph which is simply a plot of flow rate against time for a particular point on a river. Figure 2.11 shows portions of annual hydrographs for three catchments of differing size. Note that the figure does not show instantaneous rates but mean daily flows, the difference being particularly important in small catchments. Graphical techniques can be applied to hydrographs which make it possible to estimate the volumes of base and quickflow making up the total runoff so that the Base Flow Index (NERC, 1980) can be calculated, i.e. the ratio of the volume of baseflow to the total volume of runoff over a period of time. Values range from almost 100 per cent for chalk catchments to less than 20 per cent in upland catchments having thin soils overlying impermeable rock.

A feature of Fig. 2.11 is that hydrographs have a characteristic time scale, related to catchment size. This is discussed further below. During dry weather, when flow is sustained by withdrawal from storage, the hydrograph tends towards an exponential curve. The flow rate is, itself, a measure of the amount of water in store and, if the rate of change of flow is proportional to the flow, an exponential curve results.

There are a number of measures of hydrograph properties which are tabulated and commonly available from national and regional organisations. The long term average flow is an obvious one and this, together with seasonal variation, expressed as monthly mean values, are essentially climatic characteristics. The other measures such as the frequency distribution of daily flows and the extreme values—floods and droughts—are dependent on catchment characteristics as well as climate. These standard measures are not necessarily the most useful for ecological purposes. Rainfall, evaporation and runoff are usually expressed as depths, in mm, over the catchment while flow rates are quoted in $m^3 s^{-1}$. Conversion factors are listed in Table 2.2.

2.2.2 Climate Classification

Given that the physical characteristics of fresh waters are determined, ultimately, by the topography and climate of the catchment, it is desirable that the influence of climate should be linked to a general

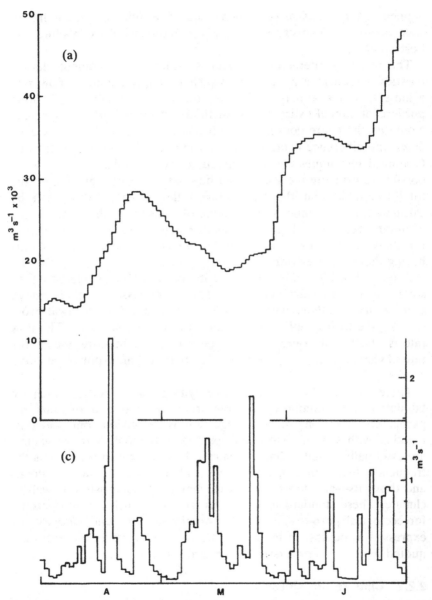

Fig. 2.11 The effect of catchment area, A_c, on flow characteristics. Figure shows typical hydrographs for the months of April, May and June. (a) River Yangtse ($A_c = 1.2 \times 10^6 \text{ km}^2$ approx); (b) River Thames ($A_c = 5000 \text{ km}^2$ approx); (c) Trout Beck ($A_c = 10 \text{ km}^2$ approx).

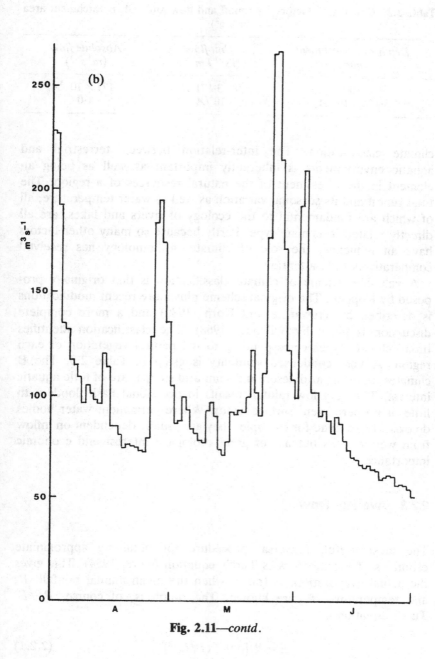

Fig. 2.11—*contd*.

Table 2.2 Conversion factors for runoff and flow rates (A_c is catchment area in km^2)

Depth over catchment (mm)	Unit flow ($l\,s^{-1}\,km^{-2}$)	Absolute flow ($m^3\,s^{-1}$)
100	3·171	$3·171 \times 10^{-3}A_c$
$3·154 \times 10^4/A_c$	$10^3/A_c$	1·0

climate classification. The inter-relation between terrestrial and aquatic environments is inherently important as well as being an element in the assessment of the natural resources of a region. The total runoff and its seasonal variation as well as water temperature, all of which are fundamental to the ecology of rivers and lakes, are all directly related to climate type. Partly because so many other factors have an influence, the role of climate in limnology has received comparatively little attention.

A valuable scheme of climate classification is that originally proposed by Koppen. The original scheme plus more recent modifications is described by Trewartha and Korn (1980) and a more complete discussion is given by Wilcock (1968). The classification identifies broad climate types corresponding to the natural vegetation of each region. A very condensed summary is given in Table 2.3. The B climates, covering arid desert and semi-arid steppe, are of little aquatic interest. The very low rainfall results in occasional flash floods with little or no permanent surface water. Where permanent water bodies do exist, Lake Chad for example, they are usually dependent on inflow from wetter areas but are of great ecological interest and economic importance.

2.2.3 Average Flow

The most useful, universal procedure for obtaining approximate estimates of average flow is Turc's equation (Turc, 1954). This gives the actual evaporation, E (mm), when the mean annual rainfall, P, and temperature, θ, are known. The runoff is, of course, $P - E$. Turc's equation is

$$E = P/[0·9 + (P/J_*)^2]^{0·5} \tag{2.2.1}$$

Table 2.3 Simplified outline of the Koppen climate classification, based on Trewartha and Korn (1980)

Climate type	Division based on rainfall	Division based on temperature	Vegetation
(A) Tropical humid climates			
AF tropical wet climate	No distinct dry season	No distinction—high temperatures-throughout year	Tropical rain forest
Aw tropical wet and dry climate	Distinct dry season when sun low (winter)		Dry woodland and savannah
(B) Dry climates			
Bw arid desert climate	Very low rainfall, distributed erratically	h—mean annual temperature >18°C	Desert
Bs semi arid steppe		k—mean annual temperature <18°C	Grassland
(C) Sub-tropical climates			
Cs Sub-tropical dry summer climate (Mediterranean)	Annual dry season—rainfall in driest month <30 mm	No distinction	Hard leaf woodland
Cf sub-tropical humid climate	No distinct dry season rainfall in driest month >30 mm		Deciduous and mixed woodland
(D) Temperate humid climates			
Do temperate oceanic climate	Rain throughout year	Average temperature of coldest month >0°C	Deciduous forest
Dc temperate continental climate	Accent on summer rain snow in winter	Average temperature of coldest month <0°C	Coniferous forest
(E) Polar climates			
ET tundra climate	No distinction	Average temperature of warmest month <10°C but >0°C	Tundra
EF perpetual frost		Temperature of all months <0°C	Permanent icecap

where

$$J_* = 300 + 25\,\theta + 0{\cdot}05\,\theta^3$$

Using climate data from Trewartha (1965) and the above equation, it is possible to generate relationships between rainfall and runoff which give rough estimates of the average flows likely to occur in different

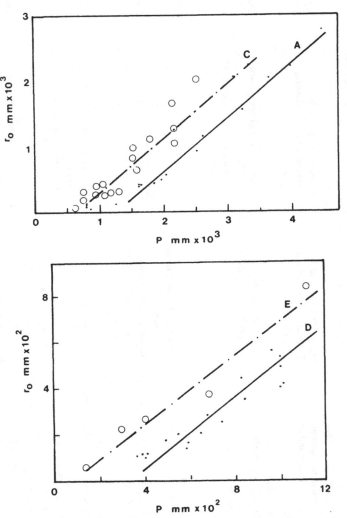

Fig. 2.12 Calculated relations between rainfall, P, and runoff, r_o, for different Koppen climate types. A climates, $r_o = 0{\cdot}82P - 1021$; C climates, $r_o = 0{\cdot}85P - 512$; D climates, $r_o = 0{\cdot}64P - 156$; E climates, $r_o = 0{\cdot}74P - 51$.

climate zones (see Fig. 2.12). At this level of detail, the runoff from B climate is, effectively, zero. The degree of scatter reflects the variability of mean annual temperature within a climate type. Although straight line relations have been fitted, there is clear indication of curvature with low mean flows. The average flow can be calculated using the conversion factors in Table 2.2.

Rainfall–runoff relations derived directly from streamflow data are obviously preferable. Figure 2.13 shows such a relation for the UK as well as the relation for rivers in Central Europe quoted by Parde (1947). Comparison with the temperate climate relation calculated above suggests that temperatures in the UK are lower than the temperate climate average.

2.2.4 Seasonal Variation in Flow

Parde (1947) published a river flow classification scheme based on the nature of the seasonal flow cycle. To make comparisons, the mean

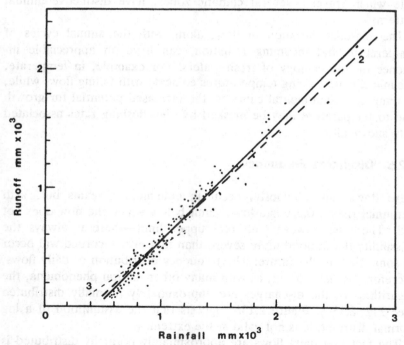

Fig. 2.13 Observed relation between rainfall, P, and runoff, r_o, in Great Britain. The equation to the full line is $r_o = 1.02P - 450$. The dashed line in Keller's equation for rivers in Central Europe ($r_o = 0.942P - 405$). The dotted line is the relation for temperate D climates (from Fig. 2.12).

monthly flows are expressed as ratios of the average annual flow. The original scheme emphasises the nature of the annual cycle—distinguishing between simple regimes where there is a single cause of high flow and complex regimes where there are several causes. The same principle can be applied to Koppen climate classes although there can be complications, particularly due to snow melt from mountainous areas (see Fig. 2.14).

Tropical and subtropical climates are characterised by high evaporative capacity throughout the year and river flow patterns reflect rainfall variability. In temperate, oceanic climates, rainfall is more or less continuous throughout the year and it is the increased summer evaporation that determines the seasonal pattern. In temperate, continental climates, runoff can be due to both snowmelt and rainfall. Here, Parde identifies a variety of flow patterns, partly reflecting altitude and the timing of snowmelt and partly the occurrence of summer and autumn rain. Major rivers, such as the Ganges and the Nile which traverse several climatic zones, have distinctive annual patterns.

The seasonal variation in flow, along with the annual cycles of temperature and incoming radiation, can have an appreciable influence on the ecology of fresh waters. For example, in temperate, oceanic climates, rising temperatures coincide with falling flows while, in temperate, continental climates, the increased potential for growth due to temperature may be masked by high flushing rates associated with snowmelt.

2.2.5 Discharge Frequency

Zero flows can, obviously, occur in ephemeral streams but, with perennial rivers, there is a lower limit below which the flow does not fall. There is, however, no real upper limit—there is always the possibility that a flood more severe than that ever recorded will occur at some time in the future. The frequency distribution of daily flows, therefore, is skewed but, as with many other natural phenomena, the logarithms of the discharges are approximately normally distributed (see Fig. 2.15a). Figure 2.15b suggests that the assumption of a log normal distribution is not valid at the extremes.

The fact that daily flows are approximately normally distributed is utilised in the standard hydrological presentation of discharge frequency. Conventional statistical practice is turned on its side in flow

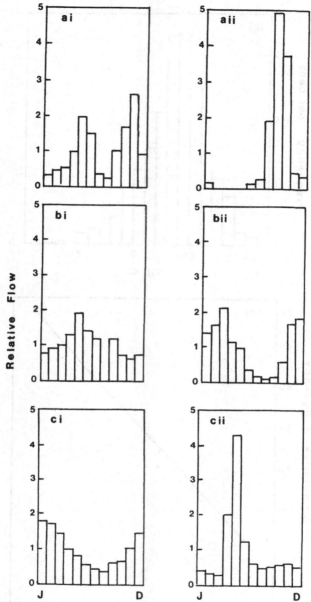

Fig. 2.14 Seasonal variation in flow in different climate zones. Mean monthly flows are expressed as ratios of the mean annual flow. (ai) tropical wet climate, Lobaye River, (Zaire); (aii) tropical wet and dry climate, Irrawaddy River, (Burma); (bi) warm, wet subtropical climate, Guadeloupe River (Texas); (bii) mediterranean climate, River Arno, (Sicily); temperate oceanic climate, River Thames, (England); temperate continental climate, River Volga (Russia).

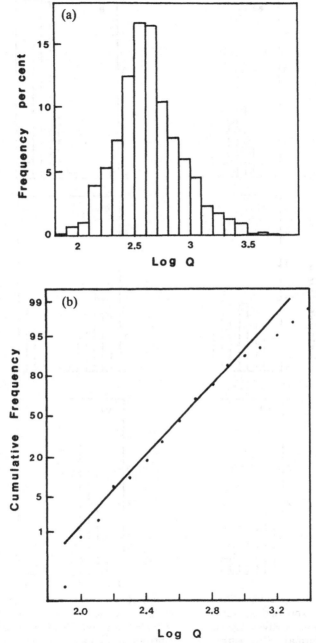

Fig. 2.15 Frequency distribution of daily flows for the River Avon, Scotland. (a) histogram of the log transformed flows; (b) same data plotted on probability paper.

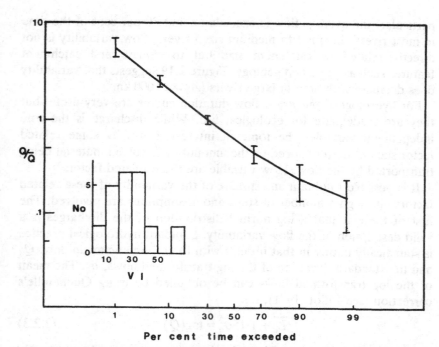

Fig. 2.16 Standardised flow duration curves for 24 British rivers, the flow, Q, being expressed as a proportion of the average flow, $\bar{Q}$. Vertical bars indicate one standard deviation on either side of the mean. The inset shows the range of values of the variability index, v_i.

duration curves and the discharge scale is often expressed as percentage of average flow (Fig. 2.16). Flow variability is indicated by the slope of the curve. Two measures of variability are used (Lord, 1966)—the statistically correct standard deviation of the logs of the discharges, σ_q, or the physically more meaningful variability index, v_i, which is the ratio of the flow exceeded 1 per cent of the time to that exceeded 90 per cent of the time. The two are, obviously, related, Lord's data for 24 British rivers producing the following relation

$$\sigma_q = 0{\cdot}289 \log v_i - 0{\cdot}027 \qquad (2.2.2)$$

Figure 2.16, again based on Lord's data, shows the average flow duration curve for British rivers derived by averaging flows expressed as a percentage of the mean flow for the different times. The mean value of the variability index is 29·0. The indications are that the general form of generalised flow duration curves has wide validity. In particular, the average flow is exceeded about 30 per cent of the time in most rivers. In small to medium sized rivers, flow variability is not directly related to catchment size but to climate and catchment features such as slope and geology. Figure 2.19 suggests that variability does decrease with area in large rivers ($A_c > 10\,000\ km^2$).

For hydrological purposes, flow duration curves are very useful but they are inadequate for ecological use. While discharge is the true independent variable, the topic of interest is usually some related factor such as current speed or the amount of dissolved material being transported by the flow. How variable are these related factors?

It is possible to obtain an estimate of the variability of these related factors although a number of steps and assumptions are involved. The first of these is that a log normal distribution of the discharges is a valid description of the flow variability. Standard hydrological practice is statistically untidy in that it deals with the arithmetic mean flow, $\bar{Q}$, and the standard deviation of the log transformed flows, σ_q. The mean of the log transformed flows can be obtained by using Quenouille's correction (see Elliot, 1971), i.e.

$$\overline{L_q} + 1{\cdot}15\sigma_q^2 = \log(\bar{Q}) \qquad (2.2.3)$$

The mean of the log transformed values, L_q, corresponds, of course, to that exceeded 50 per cent of the time. It is also assumed that the related variable, Y, is a power function of the discharge, Q, i.e.

$$Y = aQ^b \quad \text{or} \quad \log Y = \log a + b \log Q \qquad (2.2.4)$$

The final step is to make use of the statistical theorem (see, for example, Smith, 1969) which states that, if x is normally distributed with mean, μ, and variance σ^2, and $y = cx + d$, then y is normally distributed with a mean $= c\mu + d$ and variance $c^2\sigma^2$. The mean, L_y, and variance, σ_y^2, of the logs of the dependent variable can, thus, be estimated, i.e.

$$\overline{L_y} = b\overline{L_q} + \log a \qquad (2.2.5a)$$

$$\sigma_y^2 = b^2\sigma^2 \qquad (2.2.5b)$$

A duration curve for the dependent variable, similar to that for flow, can then be plotted and the arithmetic mean of the dependent variable obtained by a further application of Quenouille's correction. Some judicious drawing on the duration curve may make it possible to estimate the frequency of occurrence of the more extreme values of the dependent variable. Use of this procedure is illustrated in the numerical example on rivers (Section 5.2.4).

2.2.6 Floods and Low Flows

When a flood is caused by an isolated rainfall event, the resulting flood hydrograph has a characteristic form (Fig. 2.17). As a flood travels down river, the form of the hydrograph changes—the peak becoming less pronounced and the base wider. Associated with the travel of a flood are changes in the water surface slope. The slope of the water surface is usually greater than that of the river bed during rising floods

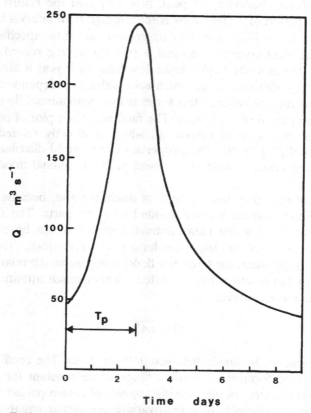

Fig. 2.17 Typical form of a flood hydrograph showing how the flow, Q, varies in time (River Spey at Boat of Garten). The time to peak, T_p, is a measure of the hydrograph characteristics.

and flatter after passage of the peak. A flood in this context simply means a high flow having these characteristics and does not necessarily imply that the surrounding land is underwater. Flood investigations make use of this characteristic form as a means of calibrating the flood behaviour of a catchment but this has little direct ecological relevance. What is important, ecologically, is the magnitude of peak flows and how frequently they occur since these have a major impact on habitats and their populations. The ecological impact of events such as floods is discussed in Section 8.3.

The relation between flood magnitude and frequency of occurrence, often referred to as extreme value analysis, is discussed in detail in the Flood Studies Report (NERC, 1975). Extreme value analysis seeks to find the relation between the peak flow rate and the return period. The return period of a given flow is the average time interval between the occurrence of flows equal to or greater than that specified. The basis of the most common analysis is that for a long record, ideally more than 25 years, the largest peak flow from each year is assumed to be a statistical distribution in which each value is independent of the others. The average value of this series is the mean annual flood which has a return period of 2·33 years. The final result is a plot of peak flow rate against the reduced variate, y, which is directly related to the return period (Fig. 2.18). The properties of the fitted distribution are such that the return period of the most probable annual flood is 1·58 years.

There are not many long records of discharge and, because of this, practical flood analysis is often divided into two parts. The first step involves determining the mean annual flood. This can be estimated from short records or from catchment characteristics. The latter procedure is the successor to earlier flood formulae which related flood flow, Q, to catchment area, A_c, often without much attention being paid to the return period, i.e.

$$Q = aA_c^b \qquad (2.2.6)$$

The exponent, b, is usually between 0·75 and 0·85. The coefficient, a, in earlier flood formulae was often taken to be constant for a region but modern practice, besides taking account of return period, replaces the constant coefficient by a multivariate regression equation which incorporates various catchment characteristics—average rainfall, soil type, extent of urban development and so on.

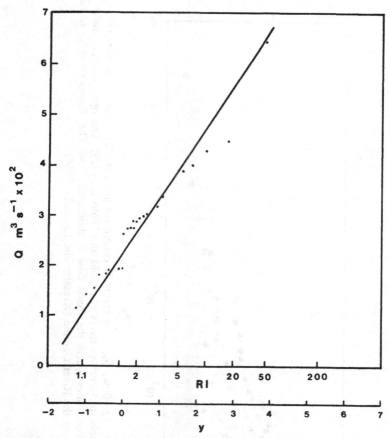

Fig. 2.18 Extreme value analysis of floods showing how the peak flow rate, Q, increases as the recurrence interval, RI, increases (River Dee above Aberdeen).

The second step utilises what long records are available. The magnitude of rarer floods are expressed as multiples of the mean annual flood and the results from different sites within a region are pooled. The resulting generalised flood curves, which show the relation between multiples of the mean annual flood and return period, are usually consistent over wide areas and can be used as a means of classifying flood behaviour.

The above procedure can be used to obtain an estimate of the bankfull discharge which is used in the estimation of channel dimen-

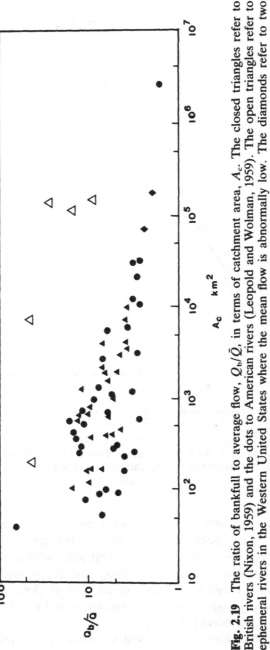

Fig. 2.19 The ratio of bankfull to average flow, $Q_b/\bar{Q}$, in terms of catchment area, A_c. The closed triangles refer to British rivers (Nixon, 1959) and the dots to American rivers (Leopold and Wolman, 1959). The open triangles refer to ephemeral rivers in the Western United States where the mean flow is abnormally low. The diamonds refer to two locations on the River Danube (Kressler and Laszloffy, 1964).

sions (see Section 5.1.3). On average, the bankfull discharge has a return period of about 1·5 years which is virtually the same as that of the mean annual flood. There is some indication that the ratio of bankfull to average flow decreases with catchment area (Fig. 2.19).

Conventional hydrological analysis leaves a gap between the use of extreme value analysis to examine floods and flow duration curves. This is understandable since modest floods that cause no structural damage are of little engineering interest and the practical use of flow duration curves is usually directed towards lower flows. This intermediate range may, nevertheless, be of ecological interest. Difficulties arise partly because flow duration curves are concerned with the features of daily flows where it has already been shown that the log normal distribution is not valid at the extremes and partly because extreme value analysis is usually based on instantaneous peak values.

Analysis of data in the Flood Studies Report (NERC, 1975) yields the following relation, for British rivers, between r_q, the ratio of the calendar day flow to the peak flow, and the S_{1085} slope measure (see

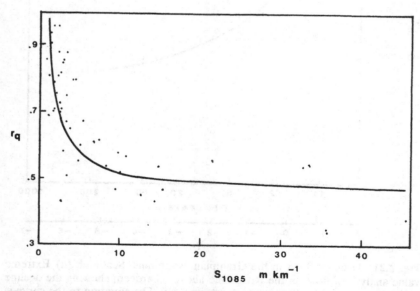

Fig. 2.20 The relation between the ratio of peak to calendar day flow, r_q, and the S_{1085} slope measure. The value of r_q is calculated from data in Table 5.1 in the Flood Studies Report (NERC, 1975) using the relation $r_q = 1/(1 + B)^N$. The curve represents eqn (2.2.7).

Table 2.4 Typical values of the flood time scale ($3T_p$)

Distance from source (L km)	Slope (S_{1085} m km^{-1})	Flood time scale ($3T_p$ days)
10	20	0·25
10^2	10	1·0
10^3	7·5	3·5
10^4	3	16[a]

[a] Extrapolating well beyond the range of eqn (2.2.8).

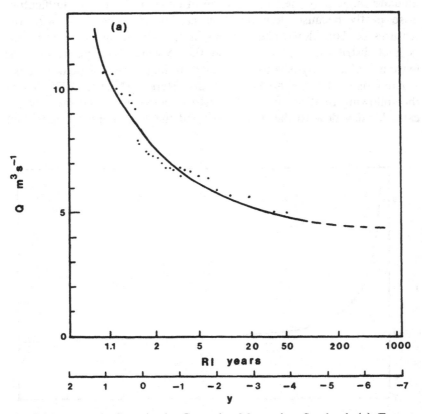

Fig. 2.21 Drought flows in the Grampian Mountains, Scotland. (a) Extreme value analysis of flows in the River Dee above Aberdeen, showing the decline in the flow rate, Q, with recurrence interval, RI. The equation to the curve is $Q = 4·21 + 3·38e^{0·52y}$; (b) the regional drought curve, i.e., the minimum recorded flow, Q, at a number of sites plotted against catchment area, A_c. The equation to the line is $Q = 1·17 \times 10^{-3}A_c^{1·153}$.

Fig. 2.20)

$$r_q = 0.608/S_{1085} + 0.46 \qquad (2.2.7)$$

The minimal annual flood, corresponding to a return period of 1·01 years, can, therefore, be converted to an equivalent daily flow and plotted on the flow duration curve as the flow exceeded 0·27 per cent of the time. This approximation goes some of the way towards closing the gap.

The number of times river flow rises and falls over the course of a year cannot be ascertained from conventional hydrological statistics. This, again, may be of ecological interest. A rough measure of the

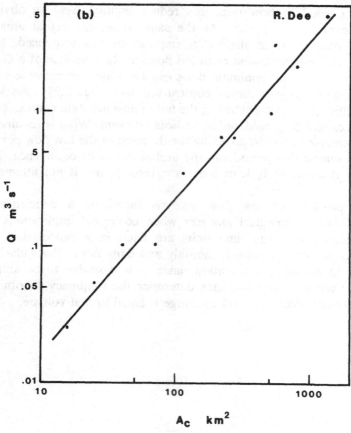

Fig. 2.21—*contd.*

number of flow reversals can be obtained by establishing a time scale for floods. Figure 2.17 suggests that the time base of a flood is about three times the time to peak, T_p. The Flood Studies Report (NERC, 1975) gives the following relation between T_p (hours) and the catchment characteristics—distance from the source, L (km), and the S_{1085} value, viz.

$$T_p = 2 \cdot 8(L/S_{1085})^{0 \cdot 47} \tag{2.2.8}$$

The correct, dimensionless measure of the number of appreciable flow reversals would be expressed as the duration of time that floods are likely divided by this time base but the former is difficult to estimate. Table 2.4 indicates some typical values of the flood time scale.

Prolonged periods of low flow which increase retention times (basin volume divided by flow rate) and reduce dilution have an obvious influence on water quality. At the same time, the actual area of habitat, particularly in shallow rivers, can be much reduced. It is possible to treat the lowest recorded flows as the converse of a flood. The series of annual minimum flows can be subject to extreme value analysis and drought curves constructed (see Fig. 2.21). As with floods, this can help in extending the tail of flow duration curves. Yet a drought cannot be considered as an isolated event. What is required is a more complex relation involving the duration of the low flow period, the flow during that period and the probability of its occurrence. This involves detailed analysis of long flow records and is not attempted here.

This problem of low flow analysis highlights a deficiency in conventional hydrological analyses when ecological implications are being considered. The time units are fixed in advance—data are analysed in terms of annual, monthly and daily flows. Particularly in relation to variation in retention times, it is desirable to be able to specify a water volume and then determine the frequency distribution of times over which the total discharge is equal to that volume.

3

Outline of Fluid Dynamics

An obvious starting point in the investigation of the relation between water movement and freshwater ecology is to examine the general characteristics of moving water. Later chapters consider the forms of water movement in specific situations and their ecological significance but all fluid motion has certain common properties.

The most obvious of these properties is that motion in the natural world is almost always turbulent, i.e., the motion is not constant but varies in speed and direction—an observation that is clear to anyone who has walked down the street on a windy day. To interpret this apparently erratic motion, it is usual to assume that it is composed of a mean motion with constant speed and direction and velocity fluctuations—gusts or eddies—which vary in both speed and direction. It is important to emphasise that the mean motion, sometimes referred to as advection, is a relative term. The mean motion to the man in the street, representative of average conditions over 1 or 2 h, contains velocity fluctuations when viewed in terms of the weather system that is generating the wind.

Some basic features of fluid motion are examined first but the bulk of the chapter is devoted to investigating the properties of turbulent motion. Given the general pattern of movement in a river or lake, it is the properties associated with turbulence—mixing, ability to maintain particles in suspension, exert a stress on the bed and the like—that have an influence on plants and animals. Because of their importance in relation to turbulence, the hydrodynamic properties of particulate matter are also discussed in this chapter.

3.1 BASIC FEATURES OF FLUID MOTION

3.1.1 Characteristics of Fluid Motion

The classic observations of the occurrence of turbulent flow were made by Osborne Reynolds over a 100 years ago. By introducing dye into a tube of flowing water, he showed that, at low velocities, the dye retained a regular, filamentous structure but that, as the velocity increased, this structure broke down into erratic swirls. The condition in which this regular structure is maintained is referred to as laminar flow.

Viscosity, the difference between oil and water and which tends to damp out turbulence, is measured by considering what happens when a horizontal force is applied to a layer of fluid. A velocity gradient occurs because, in a given time, a particle at the surface moves further than one near the base. The coefficient of absolute viscosity, μ, is defined as the ratio of the applied stress, τ, to the resulting velocity gradient, dU/dh, i.e.

$$\mu = \tau/dU/dh \qquad (3.1.1)$$

The kinematic viscosity, ν, is the absolute viscosity divided by the density, i.e. $\nu = \mu/\rho$. The units of viscosity and the relation between viscosity and temperature are considered in the Appendix.

Acceleration tends to increase turbulence—the harder the pull on an oar, the more vigorous are the eddies at the blade tip. It is logical to expect, therefore, that the criterion for the occurrence of turbulent motion will depend on the ratio of the acceleration to the viscous forces. Although it is not possible to justify it at this stage in the argument, this ratio is called, in honour of the original observer, the Reynolds Number, R_e, defined by

$$R_e = UL'/\nu \qquad (3.1.2)$$

where U is the velocity and L' is a characteristic length dimension. In rivers and lakes, this length is usually taken to be the water depth. Observations in laboratory channels show that flow is definitely laminar for $R_e < 500$ and definitely turbulent for $R_e > 2000$.

A feature of turbulent motion is that fluid elements spin about their axis as well as being translated, i.e. it is characterised by the occurrence of eddies. The strength of the rotation is measured by the vorticity, essentially the angular velocity associated with the rotations.

In laminar flow, the movement is only one of translation and the vorticity is zero. The assumption of both zero vorticity and viscosity forms the basis for ideal fluid theory. This seems completely unrealistic but ideal fluid theory is amenable to mathematical analysis and can be applied where velocity gradients are weak, for example in surface wave theory and near lake outflows.

The graphical representation of fluid motion and the resulting analysis can be envisaged as depending on whether the motion is measured with a current meter or by observing the movement of a float (Fig. 3.1). The Eulerian representation assumes the use of a

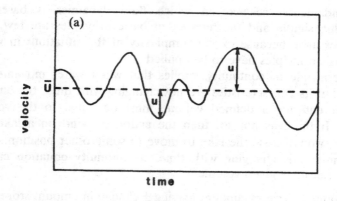

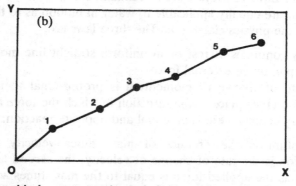

Fig. 3.1 Graphical representation of fluid motion. (a) Eulerian representation showing the velocity variation in one direction of motion at a fixed point. U is the mean velocity and u the velocity fluctuation. (b) Langrangian representation showing the position of a particle of water in two dimensions, X and Y, at successive time intervals.

current meter, i.e. it considers a fixed point in space and the motion of particles streaming past that point. In practice, the amount of detail about velocity fluctuations on the graph depends on the instrument sensitivity. The Lagrangian representation, i.e. observations with a float, considers a particular fluid particle and its position at different times. The turbulent characteristics are not displayed directly but are implicit in the different speeds and directions observed during each time interval.

3.1.2 Methods of Analysis

The fundamental science on which fluid dynamics is based is, essentially, simple and the number of basic principles are few. The difficulties arise because of the complexity of the situations in which these basic principles have to be applied.

The principle of continuity implies that when water movement is steady, i.e. its characteristics are not changing with time, the amount of water entering a defined volume must be equal to the volume leaving. If this was not so, then the action of wind on a lake, for example, would cause the lake to move to some other position. Even if the motion is changing with time, a continuity equation can be written, i.e. for a given volume

amount entering = amount leaving ± change in amount stored

Newton's Laws of Motion are so fundamental that they are re-stated here. They are equally applicable to water in motion as to the abstract bodies of the physics classroom. The three laws are:

—A body continues at rest or in uniform straight line motion unless acted upon by an external force.
—The rate of change of momentum is proportional to the applied force and takes place in the direction in which the force acts.
—To every action, there is an equal and opposite reaction.

Momentum is the product of mass times velocity and, since acceleration is the rate of change of velocity, the second law can also be stated as the applied force is equal to the mass times acceleration. By considering all the forces and accelerations acting in all three dimensions, the Navier–Stokes equations, the fundamental equations of fluid dynamics, are obtained. Unfortunately, these equations cannot be solved analytically. Either they have to be considerably simplified which simply takes one back to the application of the second law or they have to be solved numerically.

In rivers particularly, the principle of energy conservation can be applied. The total energy of a flow is composed of the potential energy due to height above some datum, the energy due to pressure within the fluid and the kinetic energy due to its motion. Bernouilli's equation states that the total energy remains constant. This is, of course, only true over short distances since, if there is motion, there will be energy losses. A basic problem in the hydrodynamics of rivers is how to estimate these losses.

The criterion for the occurrence of turbulence, Reynolds Number, has the apparent merit of being dimensionless and is only one of a host of dimensionless numbers found in fluid dynamics. There are obvious practical advantages if a critical value is independent of the units used. There is also an analytical value as it is often possible to deduce the form of relationships or critical values purely on the basis of the fundamental units involved, i.e. mass, M, length, L and time, T. In the case of Reynolds Number, it is reasonable to suppose that the occurrence of turbulence will depend on the velocity, a length dimension, i.e. the extent of the available space for eddies to be formed, as well as the properties of the fluid—density and viscosity, expressed as kinematic viscosity. Since the units of kinematic viscosity are $m^2 s^{-1}$ (see Appendix), i.e. $L^2 T^{-1}$, the only combination of velocity, length and kinematic viscosity that has no dimensions is Reynolds Number, i.e.

$$R_e = UL'/v = (LT^{-1}L)/(L^2 T^{-1}) = 1$$

3.2 THE NATURE OF TURBULENCE

3.2.1 Basic Features

An account of the basic features of turbulent motion in rivers and lakes has already been given elsewhere (Smith, 1975). Because the arguments are so central to the whole question of the ecological significance of water movement, some of the material is repeated here.

Formally, turbulent motion can be considered as having a mean motion, U, and velocity fluctuations, u, v and w in the three co-ordinate dimensions (Fig. 3.2). The velocity fluctuations are approximately normally distributed with zero mean but the root mean

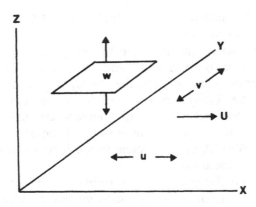

Fig. 3.2 Co-ordinate system for turbulent motion. The mean motion, U, is in the X direction and the vertical velocity fluctuations, w, generate shear stress on a horizontal plane.

square values, $\sqrt{\bar{u}^2}$ etc., are not zero. Nor are the velocity fluctuations independent. Considering the $z-x$ plane only, an upward velocity fluctuation, w, must, because of the continuity requirement, have an influence on the u velocity. The mean products, i.e. the average of the product of two simultaneous values of u and w, denoted by $\overline{uw}$, are also not equal to zero. The fluctuations of turbulent motion have a complex structure. The observations by Bowden and Fairbairn (1956) in the bottom 2 m of a tidal current are typical, i.e. $\sqrt{\bar{u}^2}/U = 0.13$ and $\sqrt{\bar{w}^2}/U = 0.66$.

Newton's second law indicates that, if there is a velocity gradient in the mean motion and there are velocity fluctuations, a shear force or stress must be generated because there is change of momentum in the fluid transported by these fluctuations. The shear stress due to the vertical velocity fluctuation, τ, is given by

$$\tau = \rho\overline{uw} \qquad (3.2.1)$$

where ρ is the density. The stress due to the lateral, cross-current fluctuations is less important.

When a viscous fluid is deformed, the shear stress is given by eqn (3.1.1). The coefficient of absolute viscosity is a property of the fluid. By analogy, the stress arising from turbulence can be expressed as

$$\tau = N \, dU/dh \qquad (3.2.2)$$

In this case, N, the coefficient of eddy viscosity, is a property of the motion and not of the fluid itself. Correctly, on the above basis, the units of eddy viscosity are $ML^{-1}T^{-1}$ and, by dimensional analysis, the stress can be interpreted as the momentum flux, i.e. the momentum transferred across unit area in unit time.

Other properties of turbulence are defined by analogy with molecular motion. The coefficient of eddy conductivity, χ, can be determined from the heat flux, J, and the temperature gradient, $d\theta/dh$, i.e.

$$J = \chi \, d\theta/dh \qquad (3.2.3)$$

Again the units of eddy viscosity are $ML^{-1}T^{-1}$. In the transport of matter, the coefficient of eddy diffusion, K, is related to the amount of matter transported across unit area in unit time and the concentration gradient, dC/dh, i.e.

$$M = K \, dC/dh \qquad (3.2.4)$$

The units of K are L^2T^{-1}. In the absence of density gradients, the three coefficients are often assumed to be roughly equal and the modern practice is to quote values of the turbulence coefficients in $m^2\,s^{-1}$. In the case of eddy viscosity, this implies writing eqn (3.2.2) as

$$\tau = \rho N \, dU/dh \qquad (3.2.5)$$

This is equivalent to assuming that N is the analogue of the kinematic viscosity, $\nu(=\mu/\rho)$. Viewing these turbulent coefficients as measures of turbulent intensity, the $m^2\,s^{-1}$ unit has the obvious merit of implying the product of length and velocity—eddy size and the velocity associated with it.

When a wind blows over a lake, energy is being continually transferred to the water yet steady conditions arise, the water is no longer accelerated and as much energy is being dissipated as received. This dissipated energy is finally converted into heat, i.e. increased molecular motion. A mechanism must exist whereby motion at the scale of the dimensions of a water body is transformed into motion at the molecular scale. Kolmogoroff (1941) and others developed this into the eddy spectrum concept. They showed that there is a complete spectrum of eddy sizes and that four main regions within it can be identified:

—the general circulation range—large scale rotational motion with dimensions of the same order as the water body and due directly to the external forces causing motion;

—the turbulent mixing range—the properties of eddies in this range are determined by the rate of dissipation of turbulent energy;

—the transition range—smaller eddies whose properties are determined by viscosity and the dissipation rate;

—viscous range—the region of molecular motion where viscosity alone is the determining factor.

The eddy spectrum gives a spatial view—all the scales of motion occurring at a particular time in a water body. An alternative form of spectral analysis can be presented—the variation in time of the scales of motion at a point—but this is less relevant in rivers and lakes where the scales of motion are restricted by boundaries. Time variation is relevant in the case of wind which is reflected in wind-induced motion in lakes. Wind has time scales of minutes associated with gusts, hours as a result of diurnal variation and days because of the passage of large scale weather systems.

In the kinetic theory of gases, the mean free path is the average distance travelled by molecules between collisions. Prandtl's mixing length model of turbulence (Prandtl, 1952) is analogous to kinetic theory. The range of sizes in the eddy spectrum is replaced by an average distance, the mixing length, l, which is the distance a fluid element travels from one level to another where it suddenly acquires the velocity of the new level.

Consider a two-dimensional flow in which there is a velocity gradient (Fig. 3.3). For a fluid element moving from level h to level

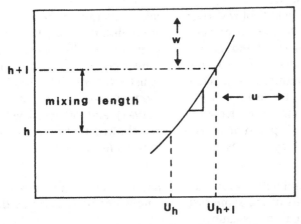

Fig. 3.3 Definition sketch for the mixing length concept.

$h + 1$, the difference between the horizontal velocities will be proportional to l and the velocity gradient, dU/dh, i.e. $U_h - U_{h+1} \doteqdot l\, dU/dh$. Prandtl also argues that the difference between the upper and lower velocities will be of the same order of magnitude as the velocity fluctuation, i.e. $U_h - U_{h+1} \doteqdot u$ so that $u \doteqdot l\, dU/dh$. The justification for the latter is the continuity requirement that the additional fluid input at the upper level requires an increase in velocity at the upper level. It is also argued that the velocity fluctuations are roughly equal so that $w \doteqdot l\, dU/dh$. Since $\tau = \rho uw$ (eqn (3.2.1)), an expression for the stress, τ, that does not involve knowing the velocity fluctuations can be written, i.e.

$$\tau \doteqdot \rho l\, dU/dh \, . \, l\, dU/dh$$

i.e.

$$\tau = \rho l^2 (dU/dh)^2 \tag{3.2.6}$$

Despite its rather tenuous derivation, eqn (3.2.6) is important in subsequent derivations. One result follows immediately. Since $\tau = \rho N\, dU/dh$ and $w \doteqdot l\, dU/dh$, we have

$$N \doteqdot wl \tag{3.2.7}$$

i.e. the eddy viscosity is proportional to the product of the vertical velocity fluctuation and the mixing length.

The nature of the turbulent motion that occurs depends on the presence and form of boundaries or discontinuities within or adjacent to the flow, i.e.

—boundary layer flow, i.e. flow in the vicinity of a rigid boundary;
—free turbulent flow where there is a distinct interface or surface of discontinuity between fluids having different velocities and/or densities;
—bulk flow where neither of the above occur.

Rivers are essentially boundary layer flows although the air–water interface is a surface of discontinuity. Careful measurement of velocity does indicate a reduction of velocity near the surface as a result of free turbulent effects. The effect is small, however, and usually neglected. All three types of motion can occur in lakes. The vertical structure of motion in a lake consists of a free turbulent layer near the surface, a zone of bulk flow often referred to as the Ekman layer and a bottom boundary layer. Additional surfaces of discontinuity can occur within the Ekman layer if the lake is stratified.

3.2.2 Boundary Layer Flow

The relation between stress and velocity gradient based on the idea of a mixing length (eqn (3.2.6)) can be written as

$$\sqrt{\tau/\rho} = l\, dU/dh \qquad (3.2.8)$$

The solution of this equation gives the velocity profile near a solid boundary. In order to solve it, two assumptions are made, i.e.

—near the boundary the stress is constant,
—the scale of turbulence is proportional to the distance from the boundary, i.e. $l = kh$ where k is Von Karman's constant.

Many experimental determinations indicate that $k = 0{\cdot}40$. $\sqrt{\tau/\rho}$ has the units of velocity and is usually referred to as the friction or shear velocity, U_f, so that

$$U_f = kh\, dU/dh \quad \text{or} \quad dU = U_f/k \,.\, dh/h \qquad (3.2.9)$$

Integrating, C and C' being integration constants

$$U = U_f/k \,.\, \ln h + C = U_f/k \,.\, \ln(h/C') \qquad (3.2.10)$$

Converting to common logarithms and taking $k = 0{\cdot}40$,

$$U = 5{\cdot}75 U_f \log(h/C') \qquad (3.2.11)$$

Equation (3.2.11) shows that $U = 0$ when $h = C'$, i.e. a definite distance away from the boundary. The anomaly is explained by the existence of a thin layer of fluid next to the boundary where the flow is laminar (Fig. 3.4). The thickness of this laminar sub-layer, δ', is given by

$$\delta' = 11{\cdot}5\nu/U_f \qquad (3.2.12)$$

The value of the constant, C', and hence the form of the profile depends on the relative dimensions of δ' and a measure of the roughness of the surface, r_p. Three states of flow can be identified, viz.

—smooth when $\delta' > r_p$,
—transitional when δ' and r_p are roughly equal,
—rough when $r_p > \delta'$.

Since δ' itself depends on the flow conditions, it follows that a given surface may be smooth under some conditions and rough under others.

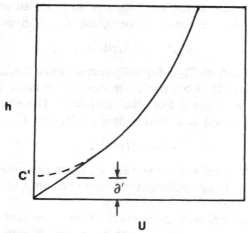

Fig. 3.4 The relation between the velocity, U, and distance from a solid boundary, h. The dashed line shows the anomalous result that arises when U is set to zero in eqn (3.2.11). δ' is the thickness of the laminar sub-layer.

For smooth flow, C' is independent of r_p and is found by experiment to be equal to $v/9U_f$ so that

$$U = 5 \cdot 75 U_f \log(9hU_f/v) \qquad (3.2.13)$$

For rough flow, $C' = 30h/r_p$ and

$$U = 5 \cdot 75 U_f \log(30h/r_p) \qquad (3.2.14)$$

This value of C' was determined (Nikuradse, 1933) from experiments with uniformly spaced sand grains. For other types of roughness, the value of r_p may not be directly related to physical dimensions. The relation of r_p to the characteristics of sediments on a river bed is discussed in Section 3.3.

The shear stress on a surface can be calculated if the velocity profile is known. An immersed body such as a fish, however, is subject to an additional force due to its form as well as the friction on its surface. This is normally expressed in terms of a drag coefficient. At a given height in a flow, the total energy of the mean motion is composed of kinetic and pressure energy and remains constant over a short distance. At the stagnation point at the nose of a static object, the velocity must be zero and the pressure will rise because the energy is constant. The drag coefficient is the ratio of the actual drag force, F_m, to the hypothetical drag force, the latter being taken as the force due to the excess pressure at the stagnation point times an area appropriate to the body, A_*, i.e. the drag coefficient, C_D, is given by

$$C_D = F_m/(A_*\rho U^2/2) \tag{3.2.15}$$

The above definition of C_D is formally correct since the kinetic energy of the flow involves $U^2/2$ but confusion can arise because some authors omit the 1/2. The above defines the total drag. The effect of friction alone can be expressed as a friction drag coefficient, C_{FD}, where

$$C_{FD} = \tau/\rho U^2/2 \tag{3.2.16}$$

Data on wind stress on a water surface are often presented in terms of a drag coefficient. Drag coefficients are not constant but vary with the nature of the motion.

The drag force opposes the motion of an immersed body. If the body is of the appropriate shape, such as an aircraft wing, some of the moving fluid is diverted so that there is a change of fluid momentum in the vertical direction. As a result, a vertical force is exerted on the body so that it may remain suspended even although it is denser than the fluid around it. Conventionally, the lift force is expressed in terms of a lift coefficient, C_L, because of the importance of the ratio of lift to drag forces in aircraft design, i.e.

$$C_L = L_f/(A_*\rho U^2/2) \tag{3.2.17}$$

where L_f is the measured lift force. Much organic matter is of such a form that lift forces may influence the rate of settlement in water.

If a particle denser than water falls through still water it will initially accelerate and then attain a constant velocity when the force causing motion due to the density difference is equal to the drag force resisting motion. This constant velocity, V_s, is variously referred to as the settling, terminal or fall velocity. For a sphere of diameter d_*, the force causing motion $= \pi d_*^3/6(\rho_s - \rho)g$ where ρ_s, and ρ are the densities of particle and water respectively. The appropriate area, A_m, for the drag force is $\pi d_*^2/4$. Equating the two forces,

$$V_s = \sqrt{1/C_D \cdot 4/3 d_*(\rho_s - \rho)/\rho \cdot g} \tag{3.2.18}$$

The settling velocity cannot be determined directly from this equation because the drag coefficient itself is a function of the particle Reynolds Number, R_{ed} ($=V_s d_*/v$). See Smith (1975) for details.

The well known Stokes Law for settling velocity is only applicable when the flow around the falling particle is laminar, i.e. when

$R_{ed} < 0\cdot5$ approximately. Stokes Law gives the settling velocity directly, i.e.

$$V_s = 1/18(\rho_s - \rho)/\rho \cdot gd_*^2 \qquad (3.2.19)$$

This is equivalent to the previous equation if $C_D = 24/R_{ed}$. McNown and Malaika (1950) introduced a modified form of Stokes Law for the laminar settling of spherical particles, i.e.

$$V_s = 1/18(\rho_s - \rho)/\rho \cdot gd^2/\phi \qquad (3.2.20)$$

ϕ is the coefficient of form resistance ($=1$ in the case of a sphere) and d becomes the appropriate length dimension. Empirical equations for the settling velocity of river sediments are given in Section 3.3.4.

3.2.3 Turbulent Structure

Vertical turbulence
The surprising thing about mixing length theory is that it seems to work. Despite the lack of rigour in its derivation, it generates theoretical results that are confirmed by observation. In particular, the logarithmic profile in a boundary layer is soundly established. In deriving the logarithmic profile, it is assumed that the stress is constant. Combining the definition of eddy viscosity (eqn (3.2.2)) with eqn (3.2.9) leads to the following expression for the vertical variation in eddy viscosity, i.e.

$$N = 0\cdot40U_f h \qquad (3.2.21)$$

In a boundary layer of finite depth, this gives a maximum value at the surface and this cannot be true. Eddy viscosity is the product of a velocity fluctuation and the mixing length and the length must be zero at a boundary. It is more likely that the stress is constant near the boundary and then decreases towards the surface—usually simplified to the assumption of a linear decrease in stress with distance from the boundary, i.e.

$$\tau = \tau_o(1 - h/D) \qquad (3.2.22)$$

Using the same equations as before, this leads to a new equation for eddy viscosity, i.e.

$$N = 0\cdot40U_f h(1 - h/D) \qquad (3.2.23)$$

The requirement that N is zero at the surface is now met and, by integrating over the vertical, the mean value, $\bar{N}$, is given by

$$\bar{N} = U_f D / 15 \qquad (3.2.24)$$

A scale for the dimensions of the mixing length can be obtained by combining eqns (3.2.2) and (3.2.22), leading to the following

$$l/D = 0{\cdot}40 h/D \, . \, \sqrt{1 - h/D} \qquad (3.2.25)$$

The mean value of l/D over the profile is $0{\cdot}107$, i.e. the mixing length is $\frac{1}{10}$ of the water depth and is independent of the velocity. Assuming $N = wl$ and the previous expression for $1/D$, the relation between the vertical velocity fluctuation and the friction velocity is obtained, viz.

$$w = U_f \sqrt{1 - h/D} \qquad (3.2.26)$$

The mean value of the ratio w/U_f is $0{\cdot}667$. The assumption is often made that the vertical velocity fluctuation is approximately the same as the friction velocity. The simple model of turbulent structure that can be derived from mixing length theory is illustrated in Fig. 3.5.

It is shown in Section 6.4 that, when a wind blows over a lake, a stress is applied at the surface and that any stress on the bed is normally negligible. In terms of mixing length theory, an isothermal lake can be viewed as an inverted boundary layer (Fig. 3.6). As for the normal boundary layer, the linear decrease in stress is an over-simplification which does not take into account the existence of a

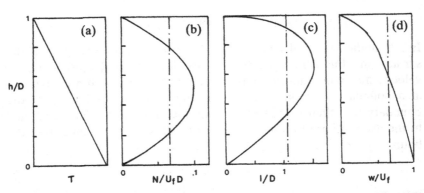

Fig. 3.5 Vertical turbulent structure in terms of the relative depth, h/D. (a) stress, τ; (b) relative eddy viscosity, $N/U_f D$; (c) relative mixing length, l/D; (d) ratio of the vertical velocity fluctuation to the friction velocity, w/U_f. Broken lines show mean values over the depth.

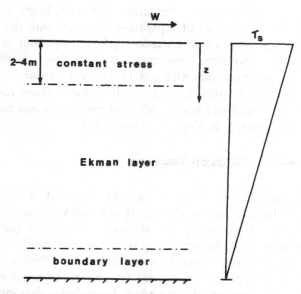

Fig. 3.6 Vertical structure in a homogeneous lake.

constant stress layer immediately below the surface. By measuring downwards from the surface, similar equations defining turbulent structure can be derived.

Longitudinal and lateral turbulence

There are longitudinal and lateral velocity fluctuations so that, theoretically, an analysis similar to that for vertical turbulence could be made. In natural, as opposed to laboratory, conditions this is not worthwhile because factors other than velocity fluctuations are more important. In most natural flows there are lateral velocity gradients—a river flows faster at the centre than near the banks. Even without velocity fluctuations, therefore, an introduced tracer will be spread longitudinally by these velocity differences and this effect may be an order of magnitude greater than that due directly to turbulence. This is usually referred to as dispersion as opposed to turbulent diffusion and measured by a dispersion coefficient but the terminology used is not always consistent.

There are further effects if a spectrum of eddy sizes exist. Two floats, initially close together, will, at first, only be moved apart by the smaller eddies as the effect of the larger will be to transport the pair.

As the distance apart increases, the larger eddies will begin to increase the separation, i.e. the rate of separation increases with the distance apart. This is the basis of Richardson's (1926) neighbour separation theory which, however, presents considerable mathematical difficulties. In practice, this effect is taken into account by having turbulence coefficients that increase with distance from the origin. Values of longitudinal and lateral diffusion and dispersion coefficients are discussed further in Sections 5.3.4 and 6.8.2.

3.2.4 Advection–Diffusion Equation

The advection–diffusion equation is the basis of what may be described as classical mixing theory. It continues the already established principle of extending the theories of classical physics, developed for motion at the molecular scale, to turbulent motion. The equation is derived from Fick's Law for molecular diffusion, extended to consider what happens when there is mean motion (advection) as well as turbulent diffusion. It has already been shown that longitudinal mixing is determined largely by dispersion and the separating effect of varying eddy sizes. The assumption of a constant diffusion coefficient is obviously questionable. Despite this reservation, the equation provides insight into the nature of mixing against which other treatments, considered later, can be compared.

The equation is fully discussed by Fischer *et al.* (1979) and only an outline of the one-dimensional case is given here. Fick's Law states that the flux of matter, M, the amount of matter transferred across unit area in unit time, is proportional to the concentration gradient, $\partial C / \partial x$, and the nature of the diffusing substance, as expressed by the molecular diffusion coefficient, D', i.e.

$$M = D' \, \partial C / \partial x \qquad (3.2.27)$$

From a mass balance equating the difference between amounts entering and leaving an element to the change of amount within the element, the diffusion equation can be written in terms of the turbulent diffusion coefficient, K, viz.

$$\partial C / \partial t = K \, \partial^2 C / \partial x^2 \qquad (3.2.28)$$

For a sudden injection of tracer into a turbulent flow, the solution to eqn (3.2.28) gives the spread of tracer concentration in terms of distance and time, $C(x, t)$, i.e.

$$C(x, t) = C_o/\sqrt{4\pi K t} \cdot e^{-x^2/4Kt} \qquad (3.2.29)$$

This solution can be compared to the equation for the normal distribution, viz.

$$y = 1/\sigma\sqrt{2\pi} \cdot e^{-(x-\mu)/2\sigma^2} \qquad (3.2.30)$$

i.e. the relative concentration, C/C_o, is normally distributed with mean, μ, $=0$ and standard deviation, σ, $=\sqrt{2Kt}$, i.e. the standard deviation increases with time (see Fig. 3.7).

If there is a velocity, U, in the x direction, then the time to move a distance ∂x, ∂t, is equal to $\partial x/U$. The change in concentration due to this translation alone, $\partial C/\partial t$, is given by

$$\partial C/\partial t = \partial C/\partial x/U = U \, \partial C/\partial x \qquad (3.2.31)$$

The combined effect of diffusion and advection, therefore, is

$$\partial C/\partial t = K \, \partial^2 C/\partial x^2 + U \, \partial C/\partial x \qquad (3.2.32)$$

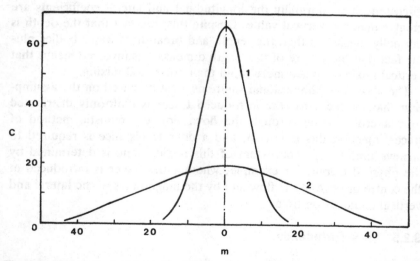

Fig. 3.7 The spread of introduced tracer in one dimension in the absence of advection. C is the relative concentration of tracer and the horizontal axis is the distance from the point of injection. The value of the diffusion coefficient is taken as $5 \times 10^{-3} \, \text{m}^2 \, \text{s}^{-1}$. Line 1 shows the spread after 1 hour and line 2 after 12 hours.

The solution for a sudden injection of tracer is, in this case,

$$C(x, t) = C_0 \sqrt{4\pi Kt} \cdot e^{-(x - Ut)^2/4Kt}$$ (3.2.33)

Again this can be compared to the equation for the normal distribution. The standard deviation of the relative concentration is the same as for diffusion alone, i.e. $\sigma = \sqrt{2Kt}$, but the mean, μ, $= Ut$, i.e. the tracer cloud moves at the same velocity as the mean motion.

O'Loughlin and Bowmer (1975) give a full discussion of the solution to eqn (3.2.32) when the tracer is injected continuously. The mathematical symbols on the page look rather intimidating although what actually happens is quite simple—the concentration ultimately becomes constant. The same authors also consider the case where the tracer is decomposing according to a first order chemical reaction as well as when the tracer is injected over a finite period.

Dimensional analysis shows that the time scale for turbulent mixing, the equalisation time, T_e, must, in terms of a characteristic length dimension, L', be given by

$$T_e = L'^2/K$$ (3.2.34)

Numerical values of dispersion coefficients in rivers and lakes are discussed later—normally the longitudinal and lateral coefficients are greater than the vertical value. Despite this, the fact that the depth is normally much less than the length and breadth of water bodies plus the fact that the square of the length dimension is involved means that vertical mixing is much more rapid than horizontal mixing.

The above one-dimensional solutions must be based on the assumption that, at the start, the introduced tracer is uniformly distributed over a cross-section through the flow. For any realistic method of tracer injection this is not so, and a definite distance is required to achieve uniformity. The extent of this mixing zone is determined by the physical layout, for example, whether the tracer is introduced at the centre or edge of the flow and by the magnitudes of the lateral and vertical equalisation times.

3.2.5 Free Turbulence

The most obvious example of free turbulence, the motion near an interface between fluid layers having different velocities and/or densities, occurs on a lake surface—fast light air over slower-moving, denser water. The most obvious features of such motion are the formation of waves and that such waves break, the crests becoming unstable and toppling over. Essentially the same happens at an interface within a water mass. Waves develop, become progressively distorted and eventually collapse (see Fig. 3.8). The explosive collapse of the distorted waves is referred to as Kelvin–Helmholtz instability. Stable conditions usually return with the formation of an intermediate mixed layer. Thorpe *et al.* (1977) report observations of the formation of such waves and their subsequent collapse at a density interface in Loch Ness.

Wind on a lake surface, besides generating waves, sets the lake surface water in motion. In the same way, fast flowing water can induce motion in stationary water. This can be seen in the flow past a projection in Fig. 3.9. Water in the lee of the projection is set in motion but, once this occurs, the continuity requirement must be met—water removed from behind the projection must be replaced.

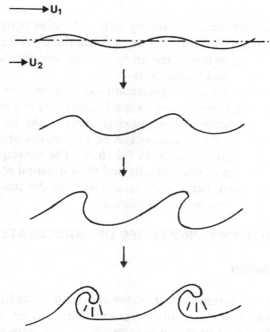

Fig. 3.8 Progressive distortion and collapse of waves on a surface of discontinuity between two currents of differing velocity (based on Mortimer, 1961).

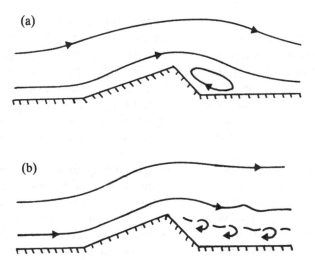

Fig. 3.9 Flow past a projection. (a) Formation of a stable rolling eddy; (b) collapse of the surface of discontinuity.

The result is the formation of a rolling eddy. At higher current speeds, instability can again occur, the surface of discontinuity collapsing to be replaced by a stream of eddies travelling with the current. Such eddies are often obvious behind bridge piers.

Motion similar to that past a projection can be seen on a large scale in lakes. The primary motion due to wind, effectively the largest scale of turbulence, can induce secondary circulation in the lee of islands and in bays. Some of these features can be seen in the observed lake surface circulation pattern shown in Fig. 6.15. The subsequent stage, collapse of the surface of discontinuity and the formation of eddies has not, as far as is known, been observed in lakes but this may be due to the absence of observers in such conditions.

3.3 HYDRODYNAMIC PROPERTIES OF PARTICULATE MATTER

3.3.1 Introduction

Turbulence theory assumes clear water and solid boundaries as in a ship's hull. This is not true of rivers and lakes. Water is usually a mixture of water and particles in suspension and only exposed bedrock is truly solid. Many boundaries in fresh water are composed of particulate matter that can be eroded and, on lake beds and in

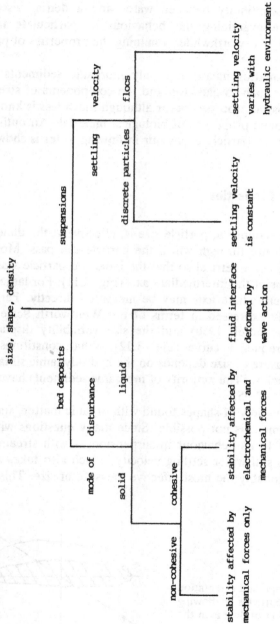

Fig. 3.10 Hydrodynamic properties of particulate matter.

slow-flowing rivers, the bed may not be solid at all. The bed is another surface of discontinuity between water and a dense, viscous fluid (mud). Before examining the behaviour of particulate matter in turbulent flow, it is worthwhile examining the properties of particulate matter itself.

Particulate matter means not only inorganic sediments but also organic matter such as plankton and the components of stream drift. The same physical processes occur although much less is known about the hydrodynamic properties of biological material. An outline classification, based on particle behaviour in flowing water is shown in Fig. 3.10.

3.3.2 Static Properties

For inorganic sediments, particle size is, effectively, the dimensions of the sieve aperture through which the particle can pass. Most natural sediments are asymmetrical so that the indicated particle size refers to the dimensions of the intermediate axis (Fig. 3.11). For large sediment sizes, the intermediate axis may be measured directly. Particle sizes are commonly expressed in terms of the Wentworth Scale (see, for example, Twenhofel, 1950) and the size variability displayed on a cumulative frequency curve (Fig. 3.12). What constitutes a representative measure of size depends on the hydrodynamic situation. This is discussed below. The majority of inorganic sediments have a specific gravity close to 2·65.

Given the variety of shapes found with organic matter, standardised size measurement is not possible. Since many questions with organic matter relate to its behaviour in suspension as with stream drift, for example, the still water settling velocity, which also takes account of density, is probably the most effective measure of size. This does not,

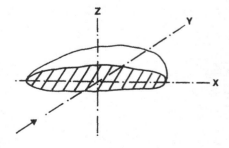

Fig. 3.11 Shape and orientation of sediment particles in flowing water. Direction of flow is in the Y direction.

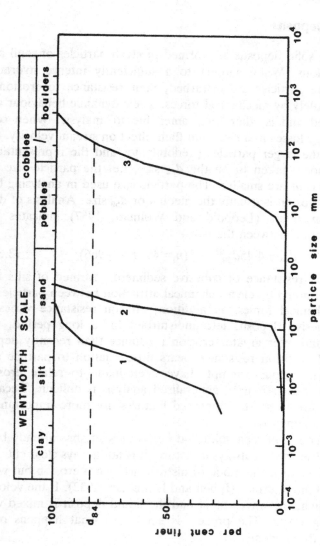

Fig. 3.12 Cumulative frequency plot of bed particle size. Samples 1 and 2 are from Loch Leven (Calvert, 1972). Sample 3 is from the middle reaches of the River Tweed.

however, completely define the properties of organic matter because, in flowing water, its form may be such that lift forces help to maintain it in suspension.

3.3.3 Bed Deposits

Non-cohesive solid deposits are formed of clean particles of sand and coarser fractions. When subject to a sufficiently intense hydraulic stress that the particles are disturbed, their resistance to erosion is determined solely by mechanical forces. Their dynamic behaviour can be generalised and is, therefore, amenable to analysis. When considering energy losses in a river and their effect on stream velocity, the influence of the larger particles predominates and the representative size is commonly taken to be the d_{84} size, i.e. the particle size for which 84 per cent are smaller. The particle size used in analysing bed stability and erosion is usually the median or d_{50} size. Analysis of data for American rivers (Leopold and Wolman, 1957) indicates the following relation between the two

$$d_{84} = 4 \cdot 45 d_{50}^{0 \cdot 83} \qquad (n = 45, \; r = 0 \cdot 966) \tag{3.3.1}$$

The erosion resistance of cohesive sediments, formed of silts and clays, is determined by electrochemical attraction between particles as well as mechanical forces. In addition, erosion resistance varies in time. A cohesive deposit, left undisturbed for a long period, will consolidate and offer greater erosion resistance than recently deposited material. Erosion resistance bears little relation to particle size and, although various attempts have been made to relate erosion resistance to shear strength, generalised analysis is difficult. Because the particles are so small, other bed features are more important in determining energy loss.

The distinction between solid and liquid deposits has already been referred to although, as always in nature, it is not always clear cut. For a true liquid deposit, the mode of disturbance is not erosion but wave formation on the interface (Ippen and Harleman, 1952). If the velocity is great enough, the waves burst and the liquid deposit is mixed with the overlying water. The process is similar to what happens on a well-defined thermocline.

3.3.4 Particles in Suspension

The theoretical basis for determining settling velocity has already been discussed. The inorganic sedimentary particles on river and lake beds are not spheres, however, and Mamak's data, quoted in Graf and Acoroglu (1966) have been used to derive settling velocity relationships for sediments (see Fig. 3.13). Three separate equations are derived depending on whether motion around the falling particle is laminar, transitional or fully turbulent (see Table 3.1). Care must be taken with the units involved. Particle sizes are normally quoted in mm but most hydrodynamic equations have to be dimensionally consistent, i.e. particle size in m if SI units are used.

The relationships in Table 3.1 are valid provided the concentration of suspended particles is not so great that the particles interfere with one another. A more serious problem in the estimation of settling velocity is likely to be flocculation. The same electrochemical attrac-

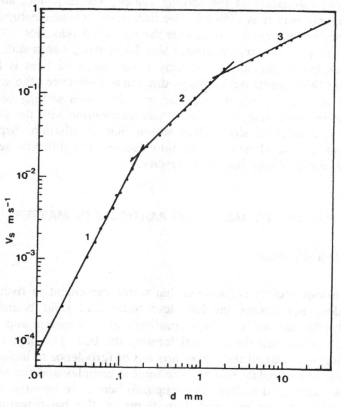

Fig. 3.13 Relation between sediment particle size, d, and settling velocity, V_s, based on Mamak's data (Graf and Acoroglu, 1966). Lines 1, 2 and 3 correspond to the equations in Table 3.1.

Table 3.1   Settling velocities for river sediments

Particle size (mm)	Settling velocity (m s⁻¹)	
	Dia. (mm)	*Dia. (m)*
$d > 1.51$	$0.134d$	$4.24d$
$0.16 < d < 1.51$	$0.109d$	$109d$
$0 < d < 0.16$	$0.683d^2$	$6.83 \times 10^5 d^2$

tion that increases erosion resistance can, if there is a weak horizontal current, cause particles to cohere and form floc blankets. Such floc blankets have a higher settling velocity than the individual particles of which they are formed. They may also trap and drag algae down with them. The estimation of the settling velocity for such flocs and for most organic matter is difficult. The difference between natural and laboratory environments may change the measured velocities.

The selection of a representative size for heterogeneous sediments in suspension is difficult, particularly if the range of sizes is large. Bagnold (1966) points out the large discrepancies between the settling velocity associated with the d_{50} size and the mean settling velocity obtained by combining the sediment size distribution with the settling velocity for different size classes within that distribution. Separate treatment of size classes, taking into account the different settling velocity–size relations, may be necessary.

3.4   THE DYNAMICS OF PARTICULATE MATTER

3.4.1   Introduction

The previous section emphasises that water movement in rivers and lakes does not always involve clear water and solid boundaries. Velocity fluctuations can keep particles in suspension and shear stresses may disturb the material forming the bed. The object here, therefore, is to extend the earlier account of turbulence to include the dynamics of particulate matter. The topic is complex and the account given is a simplified outline. The emphasis here is on how the motion of particles can be interpreted in terms of the basic features of turbulent motion rather than on quantitative accuracy. Studies of sediment behaviour in laboratory flumes are an ideal subject for doctoral theses and there is a massive literature. Valuable reviews are given by Raudkivi (1976) and Graf (1971).

The possible effect of the particles themselves on the turbulent structure is not considered. Suspended sediment may, for example, reduce velocity fluctuations although this has been disputed. The discussion is mainly concerned with non-cohesive material whose resistance to disturbance is due entirely to mechanical forces and whose concentration in suspension is sufficiently low for there to be no interaction between particles. The motion considered is a boundary layer flow having a logarithmic velocity profile as in a river. Particulate matter can also be dislodged and carried into suspension by wave action but this introduces a number of hydrodynamic difficulties because of the oscillating nature of the motion. Hydrodynamic analysis can only determine what is controlled by the flow and particle characteristics—the potential capacity for transport of particulate matter. Actual sediment loads which depend on the availability of material to be transported as well as the hydrodynamic conditions are discussed in Section 5.2.3.

A number of laboratory investigations, the earliest being those of Gilbert (1914) have shown what happens when a bed of sand is subject to increasing hydraulic stress, essentially the product of water depth and slope. The following stages are observed:

—no bed movement below a certain threshold,
—movement of isolated particles,
—whole bed in motion,
—occasional particles are carried into suspension but then fall back on to the bed, a process referred to as saltation,
—fully developed suspension.

Associated with the later stages are small scale features on the bed such as ripples. These are discussed further in Section 5.1.5.

A fundamental analytical difficulty is that, while analysis of the limiting conditions is reasonably straightforward, linking the two together is not at all easy. The limiting cases—onset of bed movement and fully developed suspension—are considered first and then possible ways of linking them are examined. In many cases, suspended matter is not derived from the bed but is either introduced from outside or generated within a water body as with plankton. The mechanics are unchanged.

3.4.2 The Onset of Bed Movement

A sediment particle is on the point of moving when the overturning moment due to the drag force is equal to the resisting moment due to the submerged weight of the particle (Fig. 3.14). Equating the two leads to the following relation between the particle size, d, and the critical stress at the onset of movement, τ_c, (Smith, 1975),

$$\tau_c = \theta(\rho_s - \rho)gd \qquad (3.4.1)$$

The densities of the particle and water are ρ_s and ρ, respectively. The value of the coefficient, θ, equal to $\tau/(\rho_s - \rho)gd$ and usually referred to as the entrainment function is discussed further below in relation to Shield's diagram.

The analysis of velocity profiles near a solid boundary emphasised the significance of the hydraulic nature of the surface as measured by the relative size of the laminar sublayer and the surface roughness. It is logical to expect, therefore, that the value of the entrainment function will be affected similarly. Comparing the particle Reynolds Number, R_{ed}, ($= U_f d/\nu$) with eqn (3.1.2) indicates that $d = \delta'$ when $R_{ed} = 11\cdot5$. Colebrook and White (1937), however, demonstrate that rough conditions occur when $R_{ed} > 3\cdot5$, and that, in such conditions, θ is constant, i.e. the critical stress increases linearly with particle size. At lower values of R_{ed} where small particles are completely submerged within the laminar sublayer, the stress required to initiate movement

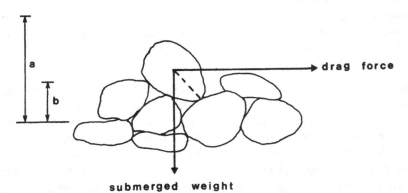

Fig. 3.14 Stability conditions for a particle on the bed. Height a indicates the thickness of the laminar sub-layer for smooth conditions ($R_{ed} < 3\cdot5$) and height b the thickness for rough conditions ($R_{ed} > 3\cdot5$).

starts to rise again so that there is a particle size for which the critical shear stress is a minimum. This can be seen when wind blows over a smooth surface—sand is blown along while fine dust is left undisturbed.

The critical value of θ refers to the onset of movement. Given the turbulent nature of the flow, it is not always clear exactly what this means. There can be a large difference in the value of θ at which movement of a few particles is observed compared to that at which general bed movement occurs. A number of attempts have been made to quantify this, i.e. to establish, for a given particle size, a relation between stress and bed load transport rate in kg s^{-1} for example. Because of the number of local features relating to sediment and channel characteristics, none of these empirical relations are satisfactory in general use. They are only valid at the site at which they were derived.

It is difficult to draw general conclusions about the stability of organic matter and invertebrates. Quite apart from behavioural responses to increasing stress in the case of invertebrates, little is known about the mechanics involved. The mode of disturbance may involve sliding rather than rotation—the breaking of the bond between some form of sucker and the stone to which an animal is attached.

3.4.3 Particles in Suspension

Analysis of particles in suspension starts by considering a form of diffusion equation, i.e. there is an equilibrium condition at which the upward transport due to turbulence is equal to the downward transport due to settling. The upward transport is the product of the coefficient of sediment diffusion, K_s, times the concentration gradient, dC/dh. The downward transport is the product of the concentration, C, times the still water settling velocity, V_s so that

$$K_s\, dC/dh + CV_s = 0 \qquad (3.4.2)$$

In order to integrate this equation, it is necessary to specify a known concentration, C_a, at some reference level, a, above the bed, i.e.

$$\int_{C=C_a}^{C=C_w} dC/C = -\int_{h=a}^{h=D} V_s/K_s\,.\,dh \qquad (3.4.3)$$

The solution for C_h, the concentration at height h, assuming that the turbulent coefficient K_s is constant and equal to N so that $K_s = U_f D/15$, is

$$C_h = C_a \exp - [(15V_s/U_f)(h - a)/D] \qquad (3.4.4)$$

(see Fig. 3.15). Equation (3.4.4) only defines the relative concentration but it is instructive to determine what fully developed suspension implies. This can be measured by the ratio of the concentration at the water surface, C_w, to C_a. If $C_w/C_a = 0.05$, then $V_s/U_f = 0.2$. Similarly, if fully developed suspension is defined as $C_w/C_a = 0.005$, then $V_s/U_f = 0.35$.

It is this need to specify the concentration at some reference level in order to solve eqn (3.4.2) that is a major stumbling block to developing a coherent theory of particle dynamics. There is no simple way of linking the reference level concentration to the earlier analysis of bed stability which is based on a critical stress. Various attempts have been made to close the gap, the first being by Einstein (1950). All are complex and none are totally satisfactory.

Central to the equilibrium conditions that are the basis for eqn (3.4.2) is that the rate of deposition on the bed is exactly matched by the rate of re-suspension. What happens if there is no re-suspension? Such conditions can arise if material is introduced from outside or with the settlement of algae. The latter is discussed by Smith (1982). The

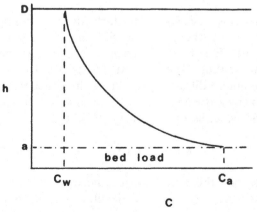

Fig. 3.15 Variation in suspended sediment concentration, C, with distance from the bed, h. C_a is the concentration at the reference level, a, and C_w is the concentration at the water surface.

basis of the mixing model used is that, at intervals, a water column is completely mixed so that the vertical distribution of matter is uniform. Between mixings, the particles settle according to still water conditions. The time scale for the process is the column clearance time, T_c, which is equal to D/V_s where D is the column depth. The analysis shows that in fully developed turbulence, corresponding to a large number of mixings within the column clearance time, the decline in the amount suspended is independent of the number of mixings such that the amount remaining in suspension, m_s, is in terms of the initial amount, m_o, given by

$$m_s = m_o e^{-t/T_c} \tag{3.4.5}$$

Some confirmation of this equation for algae is given by Gibson (1987).

Equation (3.4.5) can be derived directly from eqn (3.4.2). If there is no resuspension, there is no upward transport and $K_s \, dC/dh = 0$ and, for a column of unit area, $C = m_s/D$. Since equilibrium conditions no longer apply, we can write

$$dm_s/dt = -m_s V_s/D \tag{3.4.6}$$

Solution of this leads directly to eqn (3.4.5). The amount deposited up to time t, m_d, is, of course, the original amount less the amount remaining in suspension, i.e.

$$m_d = m_o(1 - e^{-t/T_c}) \tag{3.4.7}$$

To determine how deposited material is distributed spatially in the presence of a current of velocity U, presents a number of difficulties. Suspended sediment is subject to the same processes of dispersion and mixing as the idealised tracer considered in Chapter 4. If, however, the settling velocity of the particles is large, most of the deposition will have occurred before there is appreciable dispersion. If mixing is neglected, i.e. plug flow in the terminology of Chapter 4, the distribution downstream of the point where a 'plug' of suspended matter is introduced can be calculated directly.

For a flowing current, the elapsed time t can be replaced by x/U where x is the distance from the point of injection. The rate of deposition is the same as the change in the amount in suspension, dm_s/dt, i.e. utilising eqn (3.4.6) and eqn (3.4.5)

$$dm_s/dt = m_o e^{-x/T_c U} \cdot V_s/D \tag{3.4.8}$$

The time for the plug of suspended matter to travel $1\,m = 1/U$. The total amount deposited per unit length of channel, m_L, therefore, is given by

$$m_L = m_o/T_c U \cdot e^{-x/T_c U} \qquad (3.4.9)$$

3.4.4 Shields' Diagram

The diagram published by Shields (1936) is a plot of the critical entrainment function for the onset of bed movement, θ_c, against the particle Reynolds Number, R_{ed}. In its original form, it illustrated the then available data on the importance of R_{ed}. At high values, the entrainment function is constant but increases at low values where small particles shelter in the laminar sublayer. Superimposed on the curve are various comments about the nature of the bed and the occurrence of material in suspension. The accuracy of the original diagram can be enhanced as well as being used to display both bed stability and sediment suspension criteria in terms of the same variables.

Bridge (1981), taking more recent data into account, gives equations for θ_c that replace the original freehand curve, viz.

$$R_{ed} < 1, \qquad \theta_c = 0\cdot1 R_{ed}^{-0.3} \qquad (3.4.10a)$$

$$1 < R_{ed} < 60, \qquad \ln \theta_c = -2\cdot26 - 0\cdot905 \ln(R_{ed}) + 0\cdot168 \ln(R_{ed})^2 \qquad (3.4.10b)$$

$$R_{ed} > 60 \qquad \theta_c = 0\cdot045 \qquad (3.4.10c)$$

For inorganic sediments with a specific gravity of 2·65, a direct relation between entrainment function and particle size is more convenient. This is the form shown in Fig. 3.16. Assuming a temperature of 15°C at which the kinematic viscosity, v, $= 1\cdot146 \times 10^{-6}\,m^2\,s^{-1}$, manipulation of the original Shields variables leads to the following

$$d = [(2\cdot85 \times 10^{-7} R_{ed})/\sqrt{\theta_c}]^{2/3} \qquad (3.4.11)$$

Substituting values of R_{ed} of 1 and 60 into eqn (3.4.11) gives the limiting particle sizes, viz. 0·092 and 1·86 mm respectively. There is a discrepancy between this limit for coarse materials and that for turbulent settling (1·51 mm).

Bagnold (1966) points out that, for sediment suspension to occur, at least some of the turbulent eddies must have vertical velocity

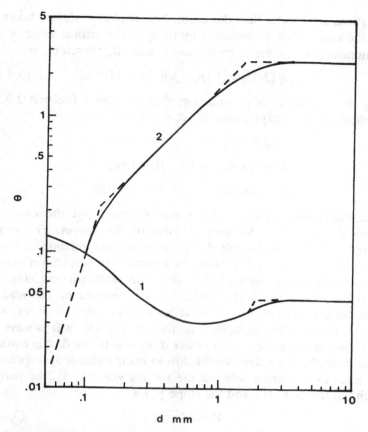

Fig. 3.16 Relation between sediment particle size, d, and the entrainment function, θ. Line 1 shows the critical value for the onset of motion at the bed (eqn 3.4.10) and line 2 the value for fully developed suspension (3.4.12).

components exceeding the settling velocity of the particles. Such an argument provides a means of expressing criteria for sediment suspension in terms of the modified Shields diagram. It has already been established that the vertical velocity fluctuations, w, are of the same order as the friction velocity, U_f, specifically that $w = 0.67 U_f$ (Section 3.2.3). There is some discussion about the relation between w or V_s and U_f (Bridge, 1981). The earlier discussion of particles in suspension indicated that, for some material to reach the surface, V_s/U_f should be no more than 0.35. This, however, neglects the turbulent characteristics of the flow and that some of the velocity fluctuations are greater than the mean. For simplicity, V_s/U_f is taken as 0·67, the same value as proposed by Bridge. The critical value of the entrainment function for sediment suspension, θ_s, therefore, is

$$\theta_s = (2\cdot25\rho V_s^2)/(\rho_s - \rho)gd = 0\cdot14V_s^2/d \qquad (3.4.12)$$

Using the settling velocity equations derived earlier (Section 3.3.4), the following direct expressions for θ_s are obtained:

$$d > 1\cdot51 \qquad \theta_s = 2\cdot5$$

$$0\cdot16 < d < 1 > 51 \qquad \theta_s = 1\cdot66d$$

$$d < 0\cdot16 \qquad \theta_s = 65\cdot31d^3$$

The significance of Fig. 3.16 is that it shows not the exact the numerical values but the general form of the curves. For coarse material, what could be called the commonsense relation is true—greater stress is required to maintain a particle in suspension than to cause movement on the bed. For very fine sand, silt and clay, the converse is true. Bagnold (1962), in fact, introduces the idea of self-sustaining suspensions—what is sometimes referred to as wash load in rivers. The argument is that a particle will always be maintained in suspension if the power it imparts to the flow is equal or greater than the work done by the flow in maintaining it in suspension. This leads to a relation between the settling velocity, V_s, the particle velocity in the flow, U', and the slope γ, i.e.

$$V_s = U' \sin \gamma \qquad (3.4.13)$$

The sediment velocity, U', is of the same order as the mean current velocity and γ is, essentially, the standard slope measure. For example, a silt particle of diameter 0·01 mm will be self-sustaining until, in a channel of slope 0·1 m km^{-1}, the velocity falls to 0·68 m s^{-1}. At any combination of slope and velocity, there is a limiting particle size that plays no part in the exchange between water and bed. With sands, the difference between the threshold values is small and this may explain, in a very general way, why specific features such as ripples occur in sand bed rivers.

3.4.5 Cohesive Sediments

The difficulties in analysing the behaviour of cohesive sediments has already been referred to in Section 3.3. A valuable review of the problem is given by Terwindt (1977). An unusual feature of the behaviour of cohesive sediments is the account of deposition in flowing water given by Metha and Partheniades (1975). Below a certain critical stress, muddy sediments settle out in the same way as non-cohesive particles but, once this critical stress is exceeded, what happens is rather unexpected. After initial, rapid deposition, the sediment concentration becomes constant and there is no further exchange between the bed and overlying water. This equilibrium concentration depends not only on the difference between actual and critical stress—which is what would be expected—but also on the initial concentration. Explanations of this are still speculative.

4

The Dynamics of Hydraulic Systems

The principle of the conservation of mass is fundamental to all physical systems. Accounting for what enters and leaves a river basin plus the storage changes and reactions within it determine many of its physical and biological features. This chapter is concerned with how such a mass balance can be calculated. Storage and reactions in a water body depend not only on the amount entering but also on how long water resides within it. This, in turn, depends on the size of the water body and the nature of the mixing that occurs. Many of the features of mixing cannot be described in hydrodynamic terms and alternative methods must be used.

4.1 LINEAR SYSTEMS THEORY

4.1.1 Introduction

Linear systems theory, originally developed for engineering use, provides a useful method of analysing some features of water movement within river basins. Here, the system is a water body or a linked series of several basins with specified inflow and transfer rates between basins while the input is an injection of tracer—a hypothetical substance not involved in any reactions within the system. It is assumed that instantaneous mixing occurs, i.e. whatever quantity of water is in a water body, it is instantaneously mixed throughout so that the concentration is the same everywhere. The object of the analysis is to characterise the behaviour of the system in terms of the relation between the tracer input and output.

In many textbooks on linear systems theory, the account of what actually happens appears to be lost in a mass of mathematics. That by Schwarzenbach and Gill (1978) provides a comprehensible account of basic ideas. The starting point of any analysis is an equation relating the input and output. In this case, this is the mass balance equation for the tracer, i.e.

> Change of amount of = Rate at which tracer − Rate at which
> tracer in water body   is introduced   tracer is
>   discharged

The restriction to linear systems means that whatever differential equation controls system behaviour, it must not contain terms that are products or powers of the variable or its derivatives. Given this restriction, the principle of superposition applies. This simply means that it is legitimate to add together the responses to different inputs—there are no interactions between the different responses.

For a single basin, the equation for the mass balance of tracer is usually quite simple and easily solved by classical methods. The previous statement of the mass balance is expressed as a differential equation in terms of the volume of the water body, V, the throughflow rate, Q, the input concentration, C_q, and the variable basin concentration, C, i.e.

$$V \, dC/dt = QC_q - QC \qquad (4.1.1)$$

Given the initial condition describing concentration in the basin at time $t = 0$, eqn (4.1.1) can be solved for a variety of conditions.

The use of Laplace transforms and transfer functions, discussed below, can be justified on several grounds. The technique can be extended to linked series of basins that are difficult to treat by other methods and provides a systematic approach to more complex situations. The method does, however, impose the additional restriction that the initial conditions are zero, i.e. there is no transient response due to the existence of tracer within the system at the start. This is not so severe a restriction as first appears. By applying the superposition principle, what happens to tracer already within the system can be considered separately from the response to added tracer and the two responses combined.

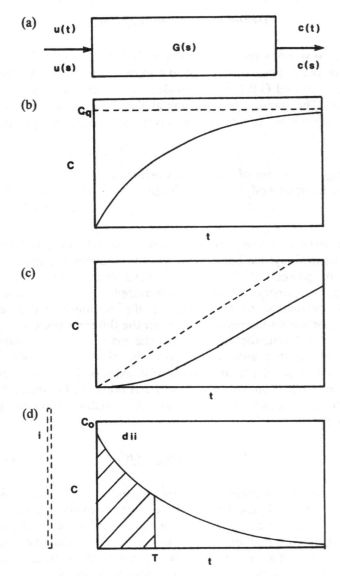

Fig. 4.1 The response, i.e. the change in concentration, C, with time, t, for various standard inputs. (a) System definition in Laplace transform terms. (b) Step function: dashed line shows the sudden increase in input concentration and the full line is the output concentration. (c) Ramp function; dashed line shows the steady increase in input concentration and the full line is the output concentration. (d) Impulse function: di represents the pulse corresponding to a sudden injection of tracer and dii the declining output concentration. The hatched area is the proportion of the amount injected that has a residence time less than T.

4.1.2 Response to Standard Inputs

The input, $u(t)$, and output, $c(t)$, of tracer are, in everyday experience, functions of time. The mathematical technique of using Laplace transforms replaces such variables in the time domain by their analogues in the frequency domain, with input $u(s)$ and output, $c(s)$ (see Fig. 4.1a). The use of Laplace transforms converts a differential equation into an algebraic one involving the variable s and whose solution can be transformed back onto the physically more comprehensible time domain.

There is a further justification. The transfer function, $G(s)$, is defined as the ratio of the Laplace transform of the output to the Laplace transform of the input, when all initial conditions are zero, i.e.

$$G(s) = c(s)/u(s) \qquad (4.1.2)$$

All mass balance equations are of the first order, i.e. they only involve the first derivative of the mass or concentration of tracer. For a first order system, the transfer function is given by

$$G(s) = 1/(1 + \tau s) \qquad (4.1.3)$$

The parameter τ is the time constant of the system. Here, the time constant is the theoretical retention time, T_r, i.e., the basin volume divided by the flow rate. The gain from all this is that, in conjunction with the superposition principle, only a few standard inputs need to be considered and whose Laplace transforms are known. The values of $c(s)$ in Table 4.1 are solutions of the equation $c(s) = G(s)u(s)$ for a first order system. The corresponding back transformations to $c(t)$ are obtained from tables.

Table 4.1 Standard input functions and their Laplace transforms

Type	Input $u(s)$	Output $c(s)$	$c(t)$
Unit step	$1/s$	$1/s(1 + \tau s)$	$1 - e^{-t/T_r}$
Unit ramp	$1/s^2$	$1/s^2(1 + \tau s)$	$t - T_r + T_r e^{-t/T_r}$
Unit impulse	1	$1/(1 + \tau s)$	$1/T_1 \cdot e^{-t/T_r}$

Step function (Fig. 4.1b)
There is an instantaneous increase in tracer concentration in the inflow corresponding, for example, to a factory coming in to operation and discharging a constant effluent. For a step of any magnitude, the output is scaled directly, i.e. if the inflow concentration is suddenly increased to C_q, the output concentration is given by

$$C = C_q(1 - e^{-t/T_r}) \qquad (4.1.4)$$

This is sometimes referred to as simple lag since the response lags behind the change in input. Ultimately, the inflow and outflow concentrations are the same.

Ramp function (Fig. 4.1c)
There is a linear increase in input tracer concentration, corresponding, for example, to the progressive deterioration in the performance of a treatment works. For concentration increasing at j_c concentration units per time, the outflow concentration is given by

$$C = j_c(t - T_r + T_r e^{-t/T_r}) \qquad (4.1.5)$$

Once a steady state is reached, the output lags behind the input, the difference between the two being $j_c T_r$.

Impulse function (Fig. 4.1d)
Mathematically, the unit impulse function is the limiting case of a pulse of unit area where the pulse duration tends to zero. In practical terms, it corresponds to the sudden injection of tracer, equivalent to the accidental discharge of a pollutant. Observing the response to a sudden injection of tracer is very useful analytically. Since the area of the pulse (Fig. 4.1di) represents the mass of tracer injected, the vertical axis has units $M\,T^{-1}$. For a pulse of magnitude f, the response is $f/T_r \cdot e^{-t/T_r}$ where $f/T_r = C_0$, i.e. the mass of tracer injected divided by the basin volume so that

$$C = C_0 e^{-t/T_r} \qquad (4.1.6)$$

The use of transfer functions requires that the initial conditions are zero but this is not a real restriction. Consider what happens when there is a sudden increase in tracer concentration in the inflow to a basin already containing tracer. The tracer already present declines in concentration as if it were an impulse response and the response to the

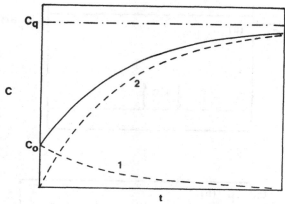

Fig. 4.2 Concentration changes as a result of a step change in input when there is tracer within the basin at the start. Line 1 shows the decline in the original tracer and Line 2 the build up of added tracer. The full line indicates the combined effect.

new concentration is independent of what is already there. The combined effect therefore, can be expressed as (Fig. 4.2)

$$C = C_0 e^{-t/T_r} + C_0(1 - e^{-t/T_r}) \qquad (4.1.7)$$

The half-life is often used as the time scale for first order systems. Here, it is, perhaps, better to think in terms of the stabilisation time, defined as the concentration to come within 5 per cent of the steady-state value. Since for $e^{-t/T_r} = 0.05$, $t/T_r = 2.995$, the stabilisation time can be taken as three times the theoretical retention time.

4.1.3 Mass Balance of a Mixed Basin

The analysis below, based on Imboden and Lerman (1978), describes the mass balance of a water body in which internal reactions take place. These reactions may be losses, for example, to the sediments, or growth. All are assumed to be first order reactions so that the net effect can be described by a single reaction rate coefficient, k'. The volume, V, the throughflow rate, Q, and the concentration in the inflow, C_q, are all constant. Instantaneous mixing occurs in the water body where the initial concentration of the substance is C_0 (see Fig. 4.3).

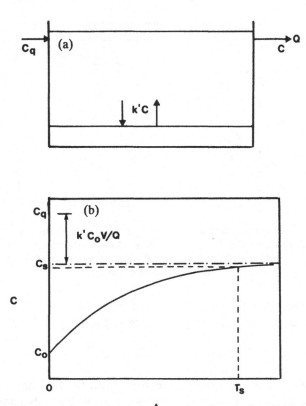

Fig. 4.3 Mass balance of a mixed basin with internal reactions. (a) Diagrammatic layout. (b) Concentration change in the basin. The curve represents eqn (4.1.8) and T_s is the stabilisation time, i.e. the time required for the basin concentration to reach 95 per cent of its steady state value, C_s.

The mass balance can be written as

$$\frac{\text{Change in amount}}{\text{per unit time}} = \text{Inflow rate} - \text{Outflow rate} + \frac{\text{Internal}}{\text{reaction rate}}$$

i.e.

$$V \, dC/dt = C_q Q - CQ + k'CV \qquad (4.1.8)$$

The solution is shown in Fig. 4.3b. The concentration at time t is given by

$$C = C_s - (C_s - C_0)e^{-(k' + 1/\tau)/t} \qquad (4.1.9)$$

where

$$C_s = C_q/(k\tau + 1) \tag{4.1.10}$$

C_s is the steady state solution when t is large and, as before, τ is the time constant or theoretical retention time. The stabilisation time is given by

$$T_s = [\ln(C_0 - C_s)/C_s] + 3]/(k' + \tau) \tag{4.1.11}$$

Besides this stabilisation time, two other time scales have already been introduced, i.e. T_r, the theoretical retention time and $1/k'$, the time scale for internal reactions within the system. There is a further time scale to be considered—the mixing or equalisation time which defines hydrodynamic conditions within the water body. If the mixing time is very short compared to the retention time, then the characteristics of the outflow will be adequately represented by assuming instantaneous mixing since introduced material will be fully mixed before any appreciable quantity has been discharged. Conversely, if the mixing time is longer than the retention time, the bulk of the inflow will be discharged before it is fully mixed and the nature of the mixing processes must be taken into account. The ratio of the retention time to the internal reaction time determines the extent to which a lake modifies water quality within a river system. If the retention time is short compared to the reaction time, the outflow water will be little different in quality from the inflow while relatively long retention times enhance the potential for water quality modification.

Once the steady state has been established, i.e. $dC/dt = 0$, $C = C_s$ and eqn (4.1.8) reduces to

$$C_s = C_q + k'C_sV/Q \tag{4.1.12}$$

If the net effect of the internal reactions is removal, then, obviously, the steady state concentration in the water body is less than the inflow concentration. How much is removed can be expressed in terms of a retention factor or coefficient, r_s, defined as the ratio of the amount removed per unit time to the total input per unit time. From eqn (4.1.12), r_s can be written as

$$r_s = k'C_sV/C_qQ \tag{4.1.13}$$

Alternatively, since $k'C_sV/Q = C_q - C_s$, r_s can also be written as

$$r_s = 1 - C_s/C_q \tag{4.1.14}$$

Both equations for r_s require knowledge of the internal reaction rate which is not always known. Where the internal reaction is due to sedimentation alone, Smith (1982) shows that, for fully developed turbulence, the sedimentation process is a first order reaction, the rate coefficient being equal to V_s/D where V_s is the still water settling velocity and D the water depth. Substitution into eqn (4.1.13) leads to the following

$$r_s = V_s C_s / (Q/A_L) C_q \qquad (4.1.15)$$

i.e. the retention coefficient is inversely proportional to the areal water load, Q/A_L, where A_L is the surface area of the lake. This agrees with the general form of the empirical equation by Kirchner and Dillon (1975) for the phosphorus retention coefficient in lakes.

4.2 INCOMPLETE MIXING

4.2.1 Mixing Processes

All discussion of tracer movement and mass balance in a water body has, up to now, been based on the assumption that complete mixing of introduced material occurs instantaneously. If the time to achieve complete mixing is short compared to the other time scales involved, this assumption may be acceptable. There are, however, situations where the assumption is clearly wrong. The object here is to examine the nature of mixing but without recourse to detailed hydrodynamic analysis. If there are internal reactions, such as plankton growth, within a water body, then how long a particle resides within it has obvious ecological importance.

An alternative assumption that can be made is that there is no mixing at all, a condition referred to as plug flow—injected tracer travels through the water body in a time equal to the theoretical retention time. Instantaneous mixing and plug flow can be regarded as the two limiting conditions. Figure 4.4 compares the outflow concentrations for these two conditions for a water body subjected to a sudden injection of tracer. In reality, the outflow concentration pattern depends on hydrodynamic conditions and basin layout such as the proximity of inflow and outflow streams. If, for example, the inflow and outflow are at opposite ends of the basin and the hydrodynamic conditions can be represented by the one-dimensional

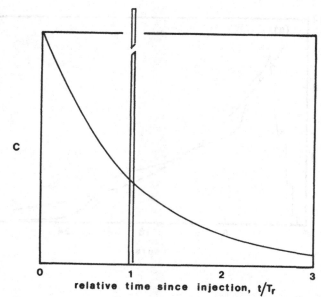

Fig. 4.4 Outflow concentration patterns following a sudden injection of tracer for limiting mixing conditions. The time scale is expressed as multiples of the theoretical retention time, T_r $(=V/Q)$. The curve shows the declining concentration assuming instantaneous, complete mixing. For plug flow, all tracer is discharged after a time equal to T_r.

advection–diffusion equation, then the outflow concentration pattern for a sudden injection would be a normal distribution (Section 3.2.4).

Direct observation of mixing characteristics based on the measurement of outflow concentration is, normally, only possible in small water bodies. Figure 4.5 shows the results of observations by White (1974) who used radioactive tracers in an experimental lake. The results predicted by the advection–diffusion equation are not substantiated and the indication is that, at low throughflow rates, conditions near to complete mixing occur while a pronounced plug flow effect can be observed at high throughflow rates.

4.2.2 Residence Time Distributions

Residence time distributions provide an alternative presentation of mixing characteristics, particularly common in chemical engineering practice. They are fully discussed by Smith (1970), for example. The time it takes for a particle to pass through a water body is termed its

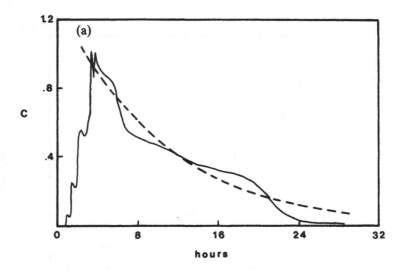

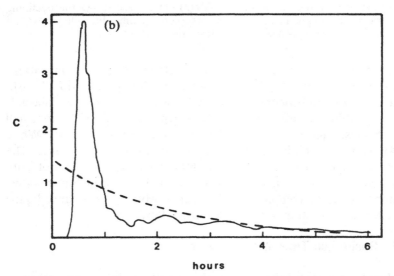

Fig. 4.5 Observed outflow concentrations from a small lake following a sudden injection of tracer. (Based on White, 1974). (a) Low throughflow rate; (b) high throughflow rate. The dashed lines are outflow concentrations assuming instantaneous mixing.

residence time, T. Except in the case of plug flow, residence time is a variable, characterised by the residence time distribution function, $J(T)$, such that $J(T)$ is the fraction of the throughflow which has a residence time less than T. Since no particle can remain in a water body indefinitely, $J(T) = 1$ as t tends to infinity (see Fig. 4.6a). The area of the hatched strip, $T\,\mathrm{d}J(T)$, represents the fraction of the throughflow which has a residence time between T and $T + \mathrm{d}T$. Since we are concerned with fractions of the total, the mean residence time can be formally defined as

$$\bar{T} = \int_0^1 T\,\mathrm{d}J(T) \tag{4.2.1}$$

The residence time distributions for the two limiting cases are easily derived. In the case of plug flow, all particles have the same residence time, equal to the theoretical retention time, V/Q. The distribution function for instantaneous mixing can be deduced from the outflow concentration curve that results from a sudden injection of tracer. Referring to Fig. 4.1d, $J(T)$ is equal to the hatched area divided by the total area under the curve, i.e.

$$J(T) = 1 - \mathrm{e}^{-T/T_r} \tag{4.2.2}$$

See Fig. 4.6b.

The residence time distribution function, $J(T)$, is the cumulative presentation of variable residence times. The alternative is the frequency curve form, obtained by differentiating $J(T)$. Referring to Fig. 4.6c, $J'(T)$ is the slope of the cumulative curve, i.e. $J'(T) = \mathrm{d}J(T)/\mathrm{d}T$, and the hatched area, $J'(T)\,\mathrm{d}T$, is the fraction of the throughflow having a residence time between T and $\mathrm{d}T$. Again a formal definition of mean residence time can be given, i.e.

$$\bar{T} = \int_0^\infty J'(T)\,\mathrm{d}T \tag{4.2.3}$$

This form has the merit of being directly related to the outflow concentration curve of a basin subject to a sudden injection of tracer. In the case of instantaneous mixing, differentiation of eqn (4.2.2) leads to

$$J'(T) = 1/T_r \cdot \mathrm{e}^{-T/T_r} \tag{4.2.4}$$

This is simply the outflow concentration curve re-scaled.

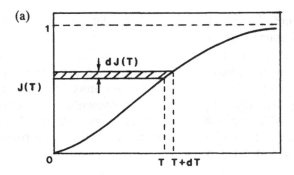

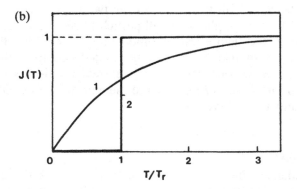

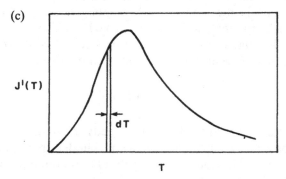

Fig. 4.6 Residence time distributions. (a) Definition sketch for residence time distribution function; (b) residence time distribution functions for limiting mixing conditions, line 1 for instantaneous mixing and line 2 for plug flow; (c) frequency form for residence time distribution.

In the limiting cases of plug flow and instantaneous mixing, the use of residence time distributions is an unnecessary formalism. Their value is in the interpretation of complex systems. The theoretical retention time, V/Q, may not be an adequate measure of how long a particle resides in a water body. Ways of deriving residence time distributions for complex systems are considered in Section 4.3.

4.2.3 Reaction Efficiency

The effect of mixing and residence time on reactions within a water body can be measured by conversion or reaction efficiency (Smith, 1970). If a substance enters a water body at a concentration C_1, leaves with a concentration C_2, and is subject to a first order reaction within the water body, i.e. $dC/dt = -k'C$, then the efficiency of the reaction can be expressed as the ratio $(C_1 - C_2)/C_1$.

For plug flow, all particles have the same residence time, T_r, so that

$$C_2 = C_1 e^{-k'T_r}, \quad \text{i.e.} \quad (C_1 - C_2)/C_1 = 1 - e^{-k'T_r} \qquad (4.2.5)$$

For instantaneous mixing, the amount in the outflow is equal to the amount in the inflow less the amount converted. This leads to the following

$$C_2 = C_1/(1 + k'T_r), \quad \text{i.e.} \quad (C_1 - C_2)/C_1 = k'T_r/(1 + k'T_r) \qquad (4.2.6)$$

The two expressions are compared in Fig. 4.7. The result may, at first sight, appear surprising—plug flow is more efficient than instantaneous mixing. With plug flow, more particles reside in the water body for longer, hence the greater conversion. At high values of $k'T_r$—fast reactions in a water body with a long retention time—both expressions converge to a value of unity, i.e. the conversion is complete and the mixing characteristics are not important. The same arguments apply to growth. Plug flow is more efficient, as pointed out by Talling and Rzoska (1967) in a discussion of Nile river plankton.

The reaction efficiency for other than the limiting cases can be calculated by making use of the frequency curve form of the residence time distribution. Each element having a residence time between T and $T + dT$ is converted such that $C_2 = C_1 e^{-k'/T}$. The total conversion, therefore, is given by

$$C_2 = C_1 \int_0^\infty J'(T) e^{-k'T} dT \qquad (4.2.7)$$

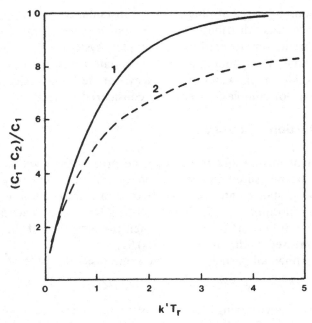

Fig. 4.7 Reaction efficiencies for limiting mixing conditions. C_1 and C_2 are the initial and final concentrations respectively. The product of the reaction rate coefficient and the theoretical retention time, $k'T_r$, is dimensionless. Line 1 represents plug flow and line 2 instantaneous mixing.

If the form of $J'(T)$ is such that the integral can be solved, an analytical solution is possible. Otherwise, an arithmetic method must be used. From the response to a sudden injection, the fractions having different residence times, and hence the amount converted, can be estimated and summed to give the total conversion.

4.3 COMPARTMENT MODELS

4.3.1 Introduction

In many cases, where the form of the residence time distribution is required, neither direct observation of injected tracer nor hydrodynamic analysis may be feasible. An alternative approach is to assume that the water body is subdivided into a number of basins with different flows or transfer rated between them and that instantaneous mixing occurs in each of the compartments. This raises questions not only about how to derive the residence time distribution for a given configuration but also about the physical interpretation of that configuration, i.e. how the layout of a water body and the hydrodynamic conditions within it can be represented by a combination of linked basins. The latter is difficult and partly a matter of judgement. Initial attention is given to the types of layout and how their residence time distributions can be derived. The basis of the analysis is to determine the response to a sudden injection of tracer. There are three basic layouts (Fig. 4.8), viz.

—systems in parallel,
—systems in series or cascades,
—systems with backflow.

These basic layouts can, of course, be combined. One arm of a parallel system, for example, may consist of a cascade of several basins. Stratified lakes can also be examined as compartment models.

4.3.2 Parallel Systems

Parallel systems are easy to analyse. If two basins with volumes V_1 and V_2 are linked in parallel such that a fraction, aQ, of the throughflow

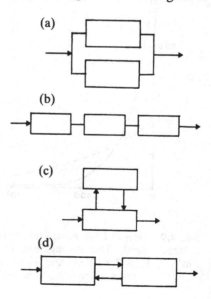

Fig. 4.8 Basic compartment model configurations. (a) Parallel system; (b) series system; (c) and (d) systems with backflow.

goes through V_1 and $(1-a)Q$ through V_2, they operate as separate single basins and the concentration in the combined outflow can be obtained by a mass balance. If m is the mount of tracer injected, the combined outflow concentration is given by

$$C = a^2 m/V_1 e^{-atQ/V_1} + (1-a)^2 m/V_2 e^{-(1-a)tQ/V_2} \qquad (4.3.1)$$

If $V_1 = V_2$ and $a = 0.5$, a result identical to that for a single basin of volume $V_1 + V_2$ is obtained. Results significantly different from that for a single combined basin only occur if the volumes and/or relative flow rates are markedly different.

Parallel layouts are more common in man-made systems such as fish farms or water treatment plants but Fig. 4.9 suggests a natural layout that could be modelled by a parallel system. The output from such a layout, also shown on Fig. 4.9 is what might have been expected. By comparison with treating the lake as a single basin, the initial outflow concentration is very much higher but falls rapidly and there are also differences at the other end. After 1000 days, there is still 0·5 per cent

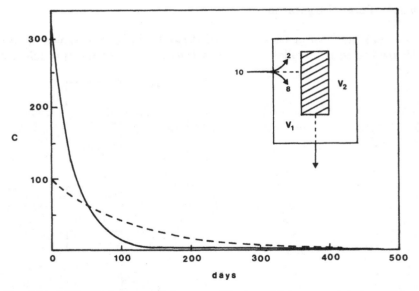

Fig. 4.9 Output from a parallel system. The inset illustrates a lake containing a large island. The throughflow is $10\,\mathrm{m^3\,s^{-1}}$, aQ is $8\,\mathrm{m^3\,s^{-1}}$, $(1-a)Q$ is $2\,\mathrm{m^3\,s^{-1}}$, V_1 is $2\times10^7\,\mathrm{m^3}$ and V_2 is $8\times10^7\,\mathrm{m^3}$. The dashed line is the output from a single basin whose volume is equal to $V_1 + V_2$.

of the initial concentration in the outflow while, for the single basin, the concentration has fallen to negligible levels (0·018 per cent).

4.3.3 Series Systems

The transfer function, $G(s)$, of a single basin has already been defined as the ratio of the Laplace transforms of the output and input when all the initial conditions are zero, i.e. $G(s) = c(s)/u(s)$. Considering two basins in series, the output from the first is the input to the second, i.e. $c_1(s) = G_1(s)u_1(s) = u_2(s)$, so that the output from the second basin, $c_2(s) = G_2(s)G_1(s)u_1(s)$. In general, therefore, a set of elements in series can be replaced by a single element whose transfer function is the product of the transfer functions of the individual elements.

For simplicity, consideration is limied to basins of equal size, i.e. all having a transfer function equal to $1/(1 + \tau s)$. Since, for a unit impulse function, $u(s) = 1$, the equation, in Laplace transform terms, for the output from a cascade of n equal basins, each having a time constant, τ, is

$$c(s) = [1/(1 + \tau s)]^n \qquad (4.3.2)$$

The solution in the time domain is

$$c(t) = (t^{n-1} \cdot e^{-t/T_r})/(T_r^n \cdot (n-1)!) \qquad (4.3.3)$$

This is the solution to a unit impulse. To obtain the response to a given input of tracer, it is easiest to consider a single basin. For $n = 1$ and an impulse of magnitude f, eqn (4.3.3) gives

$$c(t) = f/T_r \cdot e^{-t/T_r} \qquad (4.3.4)$$

Comparing this with the orignal (eqn (4.1.6)), $f/T_r = C_0$. The general solution in terms of concentration, therefore, is given by

$$C = [C_0 t^{(n-1)} e^{-t/T_r}]/[T_r^{(n-1)} \cdot (n-1)!] \qquad (4.3.5)$$

The initial concentration, C_0, is that obtained by mixing the original mass of tracer in the volume of a single basin and T_r is the retention time of a single basin. Figure 4.10 shows the effect of the number of basins on the form of the outflow concentration when the mass of tracer, throughflow rate and total basin volume are held constant.

It is possible to envisage a number of natural layouts that approximate to a cascade model—a linked series of lakes or a large lake divided into a series of basins by promontories and the like. But the

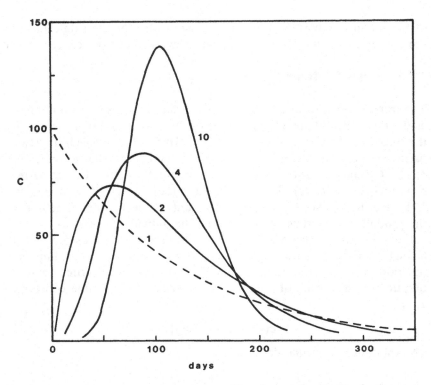

Fig. 4.10 Output from a series system. The throughflow is $10\,\mathrm{m^3\,s^{-1}}$ and the total volume is $10^8\,\mathrm{m^3}$ (as in Fig. 4.9). The numbers refer to the number of equally sized basins into which the total volume is divided.

attraction of the cascade model is its ability to reproduce a variety of observed concentration patterns. The use of the model can be reduced to a curve fitting exercise. It is quite common in chemical engineering practice to find, by trial and error, the cascade model that gives the closest fit to an observed output concentration and use the resulting equation in subsequent calculations. The output from a cascade containing between 3 and 5 elements is, commonly, close to what is observed in many systems although this may not be the correct physical interpretation. The long tails on Fig. 4.10 are often due to material being held within the water body—a feature more accurately described by the backflow or dead zone models described below.

A feature of the 10 basin cascade is that the form of the output is close to a normal distribution with the peak concentration occurring at a time equal to the retention time of the single basin—the travel time through the basin. A large number of basins reproduces, effectively, the solution to the one dimensional advection–diffusion equation (Section 3.2.4). The relation between a multi-basin cascade model and the advection–diffusion equation is discussed by Banks (1974). When the number of basins is very large, the retention time of an individual one becomes very small and the exponential term in eqn (4.3.5) declines very rapidly. Ultimately, as the number of basins becomes infinite, plug flow is recovered.

The above analytical solution only applies to a single input routed through a number of basins. In realistic, natural systems, there may be additional inputs along the cascade, in which case, numerical solutions are necessary (see, for example, Scraton, 1984). An alternative is to express the problem in terms of difference equations. This is considered in the discussion of systems with backflow.

4.3.4 Systems with Backflow

The simple system with backflow (Fig. 4.8c) represents a number of natural layouts. In a lake where a major inflow is close to the outflow, short-circuiting can occur, i.e. water travels from inflow to outflow without mixing fully with the bulk of the lake water and, in partially enclosed bays, there may be limited exchange with the main body. The system can also be taken to represent the dead zone volume in rivers caused by stagnant pools and backwaters. The assumption of instantaneous mixing, however, fails to incorporate the influence of advection and other methods are more appropriate (see Section 5.3.4).

The limiting conditions for the layout of Fig. 4.8c are elementary. If the exchange between the two basins is weak, the response tends towards that of a single basin of volume V_1. The actual response lies somewhere between these two limits (see Fig. 4.11).

Figure 4.11 represents the simplest possible layout—there may be more than two basins, some or all of which may, themselves, have inflows. Numerical solutions, therefore, are essential. An alternative to the solution of a series of differential equations is to construct difference equations, i.e. to consider changes over a series of discrete time steps. Exactly what is happening is clear, abrupt changes such as the onset of stratification in some basins can be incorporated as well as actual data on flow rates. The principle is simple and is outlined for the simple model of Fig. 4.11.

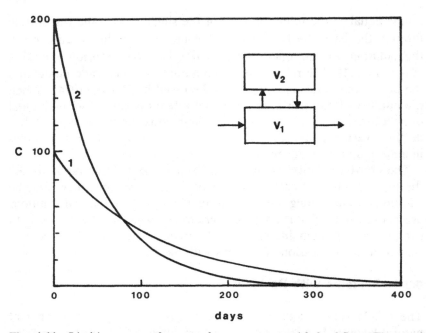

Fig. 4.11 Limiting cases of output from a system with backflow. The total volume and throughflow are the same as in previous examples and the volumes of the two basins are equal $(5 \times 10^7 \, \mathrm{m}^3)$. Line 1 represents strong exchange, the two basins acting as one. Line 2 represents weak exchange, basin 1 acting in isolation.

In terms of discrete time steps, the initial condition is included in the mass balance, i.e. over a time interval, Δt,

Amount at end = Amount at start + Amount added − Amount lost

The usual convention is that the variable concentrations are expressed as C_{ij} where i refers to the basin number and j to the time step. The balances for the first time step can be written as follows:

Basin 1 $V_1 C_{11} = C_{10} V_1 + (C_{20} + C_{21})Q_2 \Delta t/2$

$$- (C_{10} + C_{11})Q_1 \Delta t/2 \quad (4.3.6a)$$

Basin 2 $V_2 C_{21} = C_{20} V_2 + (C_{10} + C_{11})Q_2 \Delta t/2$

$$- (C_{20} + C_{21})Q_2 \Delta t/2 \quad (4.3.6b)$$

For a sudden injection of m kg of tracer into V_1, $C_{10} = m/V_1$ and $C_{20} = 0$. As the number of basins (and, hence, equations) increases, attempts to solve the set of simultaneous equations leads to clumsy algebra and, almost inevitably, error. Solution by iteration is relatively simple with a micro-computer or programmable calculator. At the first run, the initial value is assumed to be the initial value over the time step, e.g. $(C_{10} + C_{11})/2 = C_{10}$. This leads to initial estimates of C_{11} and C_{21} which are then substituted back into the equations to obtain new estimates. After 2–4 cycles, the difference between two consecutive estimates is less than a specified limit and it is possible to proceed to the second time step where the procedure is repeated.

4.3.5 Seasonal Stratification

Defining the mean retention time in a seasonally stratified lake as the total lake volume divided by the throughflow rate is not strictly correct. While stratified, the effective volume is that of the epilimnion which is not constant, the rate of thermocline deepening (length/time) being termed the entrainment velocity (see Section 6.6.4). The transfer or flow rate $(m^3 s^{-1})$ between hypolimnion and epilimnion is, therefore, equal to the entrainment velocity times the area at the depth of the thermocline. A formal analysis of this problem is given by Sangregorio (1981) but it is easier to follow what happens if it is treated as a form of compartment model and where other compartment model features can be incorporated.

A better estimate of mean residence time can be obtained by using an effective lake volume, defined as (mean volume of the epilimnion × proportion of year that lake is stratified) + (total volume of lake × proportion of year that lake is homogeneous). It is quite possible that the average flow rate during stratification is considerably different from that occurring when the lake is homogeneous. This simple correction does not reveal the complexities that can arise and it is better to consider, yet again, the response to a sudden injection of tracer.

The response depends on the time when the injection occurs. Consider first what happens with injection into a stratified lake. Given a large throughflow rate and small epilimnion volume, the tracer concentration may fall to a negligible level before overturn. During this time, the effect of entrainment is simply to increase the epilimnion volume and can probably be neglected if the mean volume over the period is used. If, however, a significant volume of tracer remains at the time of overturn, it becomes mixed throughout. Either the tracer concentration becomes negligible before the next onset of stratification or the cycle repeats itself. Equivalent cycles can be envisaged if the initial injection is in to a homogeneous lake. When stratification occurs, then tracer is present in hypolimnion and there is, as time progresses, increasing likelihood that the concentrations in the two layers will diverge because of dilution in the epilimnion. Once this happens, entrainment represents an input of tracer from the hypolimnion as well as an increase in epilimnion volume. The timing of injection becomes less important in lakes with very long retention times extending over a number of seasons.

It may be possible to dispense with a compartment model structure by making use of the superposition principle allowable with linear systems. Neglecting entrainment by considering the mean epilimnion volume over the period and assuming instantaneous mixing, the process is simply a sequence of exponential concentration declines in a

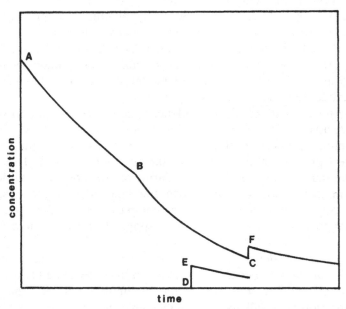

Fig. 4.12   Response to a sudden injection of tracer in a stratified lake. AB is the concentration decline during isothermal conditions. BC is the concentration decline in the epilimnion during stratified condiitons. DE is the effective input to the epilimnion due to entrainment. F is the concentration at the end of the stratified period.

single basin. During stratification, the total amount of tracer entrained is the volume of entrained water, the difference between the final and initial epilimnion volumes, times the concentration in the hypolimnion, and it may be adequate to assume that this quantity is suddenly injected into the epilimnion midway through the stratification period. The concentration at the end of the stratification period, therefore, is the sum of what remains of the concentration at the start plus what remains of the entrainment injection (see Fig. 4.12).

4.4 FINAL REMARKS

It is quite simple to develop elaborate compartment models but there is always a question in the background—are they descriptions of what is actually there or simply mathematical constructs? Chapter 6 shows that motion in lakes is complex and often unsteady so that considerable effort is required to establish the structure and transfer rates of a compartment model. The principle must be that mixing characteristics are not interesting in themselves but only in how they affect reactions in a water body. Figure 4.7 emphasises that, where fast reactions occur in a water body with a long retention time, the nature of the mixing is unimportant. Where mixing is shown to be important, it may be possible to explore the implications of different assumptions—do the resulting residence time distributions lead to significantly different reaction efficiencies? Where the problem remains, it may be possible to define more precisely the hydrodynamic questions that have to be answered. In judging the importance of mixing, differences in reaction efficiency must be set against other errors and uncertainties.

The ubiquitous, suddenly injected tracer may be a toxic substance. The analysis of mixing characteristics can also be used to determine relations between the concentration of toxin and the duration of exposure of species to that toxic concentration (see Section 8.3.4).

5

Hydraulic Characteristics of Rivers

5.1 RIVER MORPHOLOGY

5.1.1 Introduction

The object here is to provide a partially quantitative description of river morphology as a step towards understanding the nature of river habitats. In particular, it is hoped to demonstrate that there is a more consistent pattern in the structure of rivers than is sometimes imagined and that the view of some biologists that every river is different is not justified. Why rivers have certain characteristics is not considered.

A river channel is created primarily by the action of flowing water and its associated sediment load. The response of a river system to this stimulus is well expressed by following Knighton (1984) in considering the various time and length scales involved in the interaction between a river channel and its associated water and sediment inputs (see Fig. 5.1). The macro-scale features of a catchment have already been discussed in Chapter 2 and, in an ecological context, are considered to be unchanging. Features at this scale are not solely due to the action of running water since they are partly affected by other geological processes such as glaciation. This section is primarily concerned with the meso-scale features—the general form of river channels. Short term, micro-scale processes in a river associated primarily with changing rates of flow are considered in Section 5.3.

5.1.2 Classification of River Morphology

Classifying the general form and morphology of a length of river is the first step in describing the physical characteristics of river habitats. It

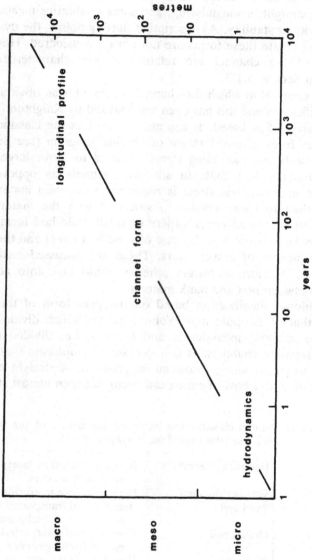

Fig. 5.1 Time and length scales in river morphology (after Knighton, 1984).

provides a generally accepted account of what a river actually looks like. The starting points for most classification schemes are the material in which the channel is formed, the plan form of the river, i.e. whether it is straight or meandering, and some qualitative measure of channel activity or stability. As descriptive starting points, the dynamic processes that create these forms are not always considered. How the dimensions of the channel are related to flow characteristics is considered in Section 5.1.3.

Using the material in which the channel is formed is an obvious first step in classification and this has been schematised by Knighton (1984) on which Table 5.1 is based. It can also be seen as the classification which merges from a consideration of Shields' diagram (see Section 3.4.4). It reflects the increasing stress required to move larger bed materials plus the fact that, in silt-clay channels as opposed to non-cohesive material, less stress is required to maintain material in suspension than to cause erosion. Associated with the material of which the bed is formed are a variety of small scale bed forms, the most obvious being ripples in the case of sand bed rivers and the pool and riffle sequences of gravel rivers. These are discussed further in Section 5.1.5. A more elaborate scheme would take into account differences between bed and bank material.

The traditional classification based on the plan form of the river channel is that of Leopold and Wolman (1957) which divides river reaches into straight, meandering and braided, i.e. divided into a number of separate channels. It is important to emphasise that these terms refer to points along a continuous gradient. A straight natural channel is rare and a host of forms can occur between almost straight

Table 5.1   River channel classification based on the nature of the bed and bank material (based on Knighton, 1984)

Cohesive	Bed rock channels	Basically limited to short, steep headwater reaches
	Silt-clay channels	High resistance to erosion
Non-cohesive	Sand bed	Bed material transported at a wide range of discharges
	Gravel bed	Coarse gravel transported at higher discharges only
	Boulder bed	Very large bed particles which are moved infrequently. Merges into bed rock channels

and fully developed meanders. A great deal of subsequent research has gone into attempting to find the combinations of discharge, slope and bed particle size associated with different channel forms.

Ferguson (1981) criticises this approach as over-simplified and suggests an alternative, based partly on earlier Russian work and which places more emphasis on features in the river and not just on the plan form on a map. Ferguson's scheme is summarised in Table 5.2. Some progress has been made in differentiating river type in terms of unit stream power—essentially the product of slope and flow divided by the channel width. This is to be expected since the degree of activity, in this sense, reflects the energy of the flow.

Ferguson's classification is valuable although a few additions, essentially the retention of some of the features of the older Leopold and Wolman scheme, may be helpful in providing this initial stage of habitat description. These are:

—a further subdivision of the active, unconfined meander class in terms of sinuosity and irregularity;
—the separation of the active, low sinuosity class into single and multi-channelled or braided types although this is not always easy and may depend on discharge at the time of observation;
—the retention of the subdivision in terms of plan form for inactive channels plus comments on the occurrence of pool and riffle sequences.

5.1.3 Relations between Channel Dimensions and Flow

Given that a river channel is created primarily by the action of water flowing within it, it is reasonable to expect there to be some relation, on average, between a measure of the flow characteristics and the channel dimensions. Any such relation is likely to vary with the material in which the channel is formed and with other morphological features.

Relations between flow and dimensions are often expressed in terms of bankfull conditions, i.e. when the flow is on the point of spilling out from its confined channel (see Fig. 5.2a). The channel section will not be adjusted to very rare floods nor will they be determined by low to average flows. It is likely, therefore, that there is an intermediate flow for which, over a period of time, the product of some measure of flow energy or sediment movement and duration is a maximum (Wolman

Table 5.2 River classification based on Ferguson (1981)

Active	Meandering { Unconfined	Rivers eroding the outside of bends and depositing point bars on the inside. No implication of exceptional sinuosity or regularity	Local imbalances in erosion and deposition can cause erratic changes in channel form. Not restricted to lowland rivers
	Confined	Normal processes of meander development are restricted by valley sides	Down valley migration is more significant than lateral development. Tendency to hug valley walls for considerable distances
	Low sinuosity {	Rivers with actively changing channels where erosion is not concentrated at bends and point bars are uncommon	Braiding often occurs with the haphazard formation of midchannel bars at times of flood
Inactive		Rivers that are not perceptibly migrating. Plan form may be sinuous, irregular or straight	Natural basic types are: —bed rock channels, —gravel rivers with erosion resistant banks, particularly tree-lined, —lowland rivers with cohesive banks, inactivity may be due to river training works

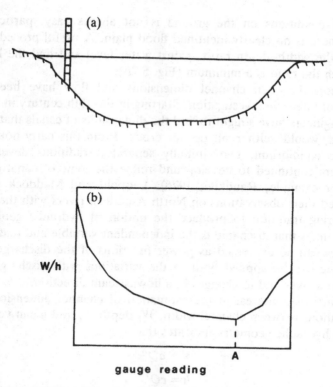

Fig. 5.2 Bankfull conditions. (a) Definition sketch. (b) Bankfull conditions occur when the width to depth ratio, W/h, is a minimum, i.e., at gauge reading A.

and Miller, 1960). This intermediate flow is often referred to as the dominant discharge.

The limited evidence that bankfull and dominant discharges are equivalent is based on the fact that they have, on average, the same frequency of occurrence. Both have recurrence intervals of about 1·5 years, i.e. for both the discharge computed by Wolman and Miller (1960) and the bankfull flows measured by Wolman and Leopold (1957) the average interval between occurrences of flows equal to or greater than this value is 1·5 years and this is, effectively, the same as the most probable annual flood (see Section 2.2.6). For individual rivers, the frequency of occurrence of bankfull flow varies considerably (Williams, 1978). Despite some uncertainty about its physical interpretation, bankfull discharge is useful in describing a river. Estimating bankfull conditions on the ground is not always easy, particularly where there is no clearly identified flood plain. A useful procedure is to plot the width–depth ratio against water level and find the height for which the ratio is a minimum (Fig. 5.2b).

Relations between channel dimensions and flow have been the subject of intensive investigation. Starting in the 19th century in India, canal engineers have sought to find the dimensions of canals that, over a period, would neither silt up nor erode. From this early notion of dynamic equilibrium, two, initially separate, traditions developed. Engineers continued to develop and refine the logic of canal design (see, for example, Raudkivi, 1976). Leopold and Maddock (1953) combined their observations on North American rivers with the canal engineering tradition to produce the notion of hydraulic geometry. This assumes that discharge is the independent variable and that other variables can be expressed as power functions of the discharge. The technique can be applied both to the variations with discharge at a point on a river and to changes in a downstream direction.

Of particular interest in the estimation of channel dimensions are the relations between channel width, W, depth, h, and mean velocity, U. The hydraulic geometry indicates that

$$W = aQ^b \tag{5.1.1}$$

$$h = cQ^f \tag{5.1.2}$$

$$U = kQ^m \tag{5.1.3}$$

The discharge, Q, is equal to WhU, so that

$$Q = aQ^b \cdot cQ^f \cdot kQ^m \tag{5.1.4}$$

It follows that

$$ack = 1 \tag{5.1.5}$$

and

$$b + f + m = 1 \tag{5.1.6}$$

The hydraulic geometry principle stimulated a flurry of research, particularly on variation in the exponents, b, f and m (see Park, 1977). Clearly, if all the exponents and coefficients were known, the dimensions of any channel at any flow could be predicted.

For the practical prediction of the approximate dimensions of a river channel, an alternative procedure, based on conditions at bankfull is outlined below (see Fig. 5.3). What is being calculated is immediately

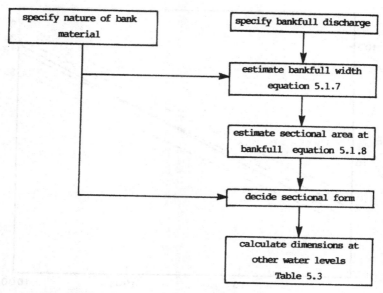

Fig. 5.3 Procedure for estimating channel dimensions.

clear and any local knowledge that is available can be taken into account. It is assumed that the bankfull discharge has already been established.

Relations between channel width and discharge are surprisingly consistent. Leopold *et al.* (1964), for example, show a relationship extending over a range of flows from less than $1\,\mathrm{m^3\,s^{-1}}$ to more than $200\,000\,\mathrm{m^3\,s^{-1}}$ (River Amazon at Obidos). The approximate rule is that the width is proportional to the square root of the discharge. The exact nature of the relation depends on the resistance of the banks to erosion. Schumm (1960) shows that increasing quantities of cohesive silt reduces the width–depth ratio and the presence of trees on the bank can reduce river width (Charlton, 1975). The results of several more detailed investigations are shown in Fig. 5.4. In general, the relation between the top water width at bankfull, W_{bx}, and the discharge, Q_b, is

$$W_{bx} = a\sqrt{Q_b} \qquad (5.1.7)$$

In the absence of any local knowledge, the approximate value of the coefficient is 3·5.

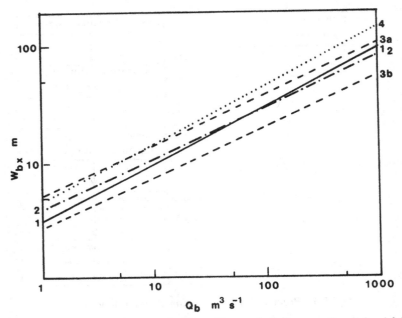

Fig. 5.4 Relation between top width, W_{bx}, and discharge, Q_b, at bankfull. The full line (1) is the suggested standard relation, i.e., $W_{bx} = 3·5Q_b$. Line 2 is from Nixon (1959), lines 3a and 3b define the range suggested by Charlton (1975) and line 4 is Lacey's relation for sand rivers.

The total cross-sectional area at bankfull, A_b, appears to be less dependent on the nature of the bank material. Williams' (1978) equation is based on data from a wide variety of climatic and lithological types. Since the scatter is surprisingly small, it can be rearranged to give an estimate of A_b, viz.

$$A_b = [(1·73Q_b)/S^{0·28}]^{0·826} \qquad (5.1.8)$$

S is the channel slope in m km^{-1}.

Other channel dimensions are dependent on the sectional form of the channel, it being commonly assumed that sections are either parabolic or trapezoidal (see Fig. 5.5). The parabolic form occurs where there is little distinction between bed and bank, both being formed of the same non-cohesive material. The trapezoidal form is more common where the bank material is more erosion resistant than the bed.

The parabolic section is generated by the equation $y = ax^2$, the

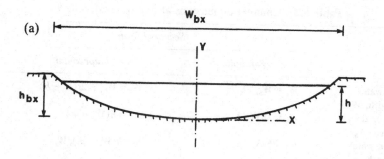

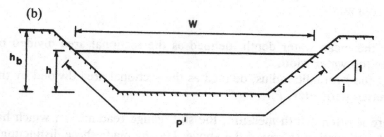

Fig. 5.5 Idealised cross-sectional form of river channels. (a) Parabolic; (b) trapezoidal.

co-ordinate system being centred on the channel bed at the centre line. Since $y = h_b$ at $x = W_{bx}/2$, $a = 4h_b/W_{bx}^2$. For the trapezoidal section, the bankside slope, j, must also be specified. For natural rivers, the side slope does not usually exceed 3 and may be steeper if the banks are erosion resistant. The maximum depth at bankfull, h_b, can be calculated from the following

Parabolic $\qquad\qquad h_b = 3A_b/2W_{bx}$ $\qquad\qquad$ (5.1.9a)

Trapezoidal $\qquad\quad h_b = (W_{bx} - \sqrt{W_{bx}^2 - 4jA_b})/2j$ $\qquad$ (5.1.9b)

The geometric relations characterising conditions at any other water level are listed in Table 5.3. Various measures of water depth are used in the calculations, viz.

—h_b, the maximum water depth at bankfull,
—h_m, the maximum water depth at any other level,

Table 5.3   Channel dimensions at any water depth, h

	Sectional form	
	Parabolic	*Trapezoidal*
Top water width, W_*	$W_x = W_{bx}\sqrt{h/h_b}$	$W_x = W_{bx} - 2j(h_b - h)$
Sectional area, A_h	$A_h = 2hW_x/3$	$A_h = h(W_x - jh)$
Mean water depth, $\bar{h}$	$\bar{h} = 2h/3$	$\bar{h} = h(W_x - jh)/W_x$
Hydraulic mean depth, h_r	$h_r = (2hW_x^2)/(3W_x^2 + 8h^2)^a$	$h_r = h(W_x - jh)/(W_{bx} - 2jh_b + 2h\sqrt{1 + j^2})$

^{a} Provided $W/h > 4$.

—$\bar{h}$, the mean water depth, defined as the sectional area divided by the top water width,
—h_r, the hydraulic radius, defined as the sectional area divided by the wetted perimeter, P_x.

There is often a fifth measure, the staff gauge reading, h', which has no direct relation to any of the above. Having made these distinctions, they are often neglected. In natural rivers, the water depth is usually small compared to the width and the hydraulic radius is taken to be the same as the mean depth so that the flow is, effectively, two dimensional. The above procedure is not the last word on channel dimensions but it does provide an indication of river size given the streamflow. The procedure is illustrated in the numerical example (Section 5.2.4).

5.1.4 Particle Size Variation along a River

The question of what controls the variation in particle size along the length of a river has been argued over for many years (see, for example, Leliavski, 1955). One argument assumes that all sizes of particle are available and that size variation is due to hydrodynamic processes. The counter argument stresses control by sediment availability and the fact that the downstream transport of sediment itself reduces particle size by abrasion. The object here is simply to examine some of the empirical evidence about an important feature of river habitats.

The classic study by Hack (1957) led to a relation between slope, S (m km^{-1}), median bed particle size, d_{50} (mm) and catchment area, A_c (km^2), viz.

$$S = 6 \cdot 0 (d_{50}/A_c)^{0 \cdot 6} \tag{5.1.10}$$

Clearly, particle size is not determined by slope alone and is influenced by some measure of river size as represented by the catchment area.

The ratio d_{50}/A_c is difficult to interpret and it is physically more meaningful to consider the ratio d/S in terms of A_c. The ratio d/S is Lokhtine's coefficient of fixation which is sometimes used as a measure of river stability (Leliavski, 1955). It is also related to Shield's function, θ (see Section 3.4.4), i.e.

$$d/S = a/\theta \tag{5.1.11}$$

where the cofficient, a, depends on the water depth and the density of the sediments.

Hack's original data have been re-analysed assuming that

$$d_{50}/S = aA_c^b \tag{5.1.12}$$

The results are indicated in Table 5.4. Also in the table are the results of analysing data obtained during an ecological survey of the Tayside Region of Scotland and, for North American rivers, from Leopold and Wolman (1957). None of the sites in Hack's data had catchment areas greater than 1000 km^2 and, for the other two data sets, sites where the catchment exceeded 1000 km^2 were omitted from the analysis. The power law relation appears to be acceptable although the accuracy is not high. There is a rough rule that the d_{50}/S value is proportional to the square root of the catchment area.

What happens if relationships derived from small catchments are extrapolated to areas greater than 1000 km^2? Data for sites where $A_c > 1000$ km^2 from Tayside (Leopold and Wolman, 1957) and for the River Danube (Kresser and Laszloffy, 1964) are plotted in Fig. 5.6 as

Table 5.4 Parameter values in the equation $d_{50}/S = aA_c^b$

	a	b	r	n
Hack (1957)	0·667	0·591	0·88	66
Tayside survey	1·95	0·469	0·83	50
Leopold and Wolman (1957)	1·30	0·415	0·65	37

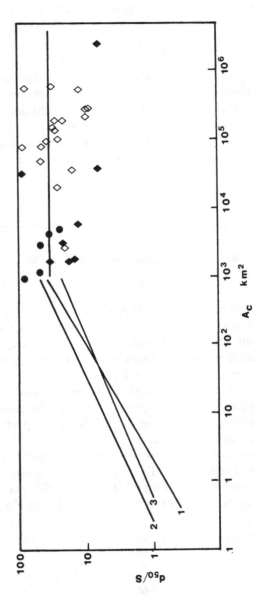

Fig. 5.6 Relations between the coefficient of fixation, d_{50}/S and catchment area, A_c. For $A_c < 100 \text{ km}^2$, line 1 is Hack's (1957) relation, line 2 is from the Tayside survey and line 3 is from Leopold and Wolman (1957). For $A_c > 1000 \text{ km}^2$, circles are Tayside data, closed diamonds are American data from Leopold and Wolman (1957) and open diamonds are for the River Danube (Kresser and Laszloffy, 1964). The horizontal line is the mean value.

are the regression lines from Table 5.4. The indication is that extrapolation of the square root relation to large rivers is wrong. The mean value of d_{50}/S for all 31 large river sites in 35·1. This compares with a calculated value based on the modified Hack equation for $A_c = 1000 \, \text{km}^2$ of 39·5. The mean large river value corresponds to a limiting area of 817 km² using the modified Hack equation.

That a relation between the coefficient of fixation and catchment area only provides a very approximate estimate of particle size variation along a river is not surprising. Indeed, the surprise is that any relation exists at all. Many factors influence the variation in particle size and these are not represented adequately by catchment area alone. Closer examination of the Danube data, for example, shows that the especially high values of d/S occur just below the confluence of major tributaries. Despite these reservations, it seems clear that the systematic increase in d/S with area in small rivers is not maintained with increasing river size.

There is the temptation to conclude that hydraulic control dominates in small rivers but that, with increasing river size and distances, the sources of sediment, the question of sediment availability, in terms of size, becomes increasingly important. The ability to compare observed and predicted values of the coefficient of fixation in both small and large rivers emphasises its value as a measure of river stability.

5.1.5 Micro-topography of River Beds

Uniform distribution of sediment over a channel bed is unusual and the occurrence of a variety of characteristic forms is much more common. Attempts to explain their occurrence in terms of the hydrodynamic processes involved have not proved conclusive. There are two basic types of bed form, viz.

—large scale bars having dimensions of the order of the channel width,
—smaller scale features occurring in sand bed rivers.

Large scale bars reflect the features of river morphology discussed earlier and have an obvious influence on the nature of river habitats. Their occurrence must be related, in some way, to the mechanisms controlling sediment movement. They have a varied particle size composition, tending towards the larger sizes, and are usually exposed at low flows. Bars can be classified according to their orientation

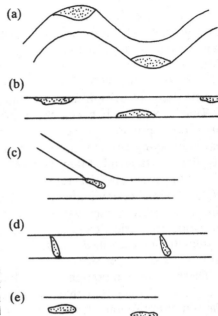

(a)

(b)

(c)

(d)

(e)

Fig. 5.7 Orientation of shingle banks (bars) in rivers. (a) Point bars; (b) alternate bars; (c) junction bars; (d) transverse bars (pools and riffles); (e) island bars.

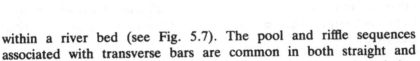

within a river bed (see Fig. 5.7). The pool and riffle sequences associated with transverse bars are common in both straight and meandering channels. A feature is their consistent spacing, being commonly 5–7 channel widths apart.

It has already been shown (Section 3.4.4) that the difference between the critical values of the entrainment function for the onset of bed movement and fully developed suspension is small in sand rivers. A high degree of mobility, therefore, is to be expected. Once bed movement is started, a sequence of forms occur which, although partly dependent on particle size, are primarily determined by the flow conditions (see Table 5.5). Discharge, effectively water depth times velocity and slope, have been combined in a number of ways by different investigators to indicate the conditions in which the various forms occur. Plane beds and, particularly, antidunes require the combination of a sand bed and a steep slope which is rare other than in the laboratory although they may occur in ephemeral, desert streams. Ripples and dunes are much more common and can be seen as a form of self-regulation. Without the additional resistance to flow

Table 5.5 Bed forms in sand bed rivers

Bed form	Characteristics	Nature of motion
Plane bed		Below threshold of particle movement
Ripples	Asymmetrical form with sharp downstream face; wavelength dependent on particle size but not on flow depth	Predominantly bed movement
Dunes	Much larger than ripples but similar in form; wavelength proportional to flow depth	Increasing proportion of particles carried into suspension
Plane bed		Dunes washed away; fully developed suspension
Antidunes	More symmetrical than dunes; occurring in steep channels	Sediment very disturbed; bed form migrates downstream

that they cause, the velocity would be so high that the channel would destroy itself.

5.2 RIVER MECHANICS

This section is concerned with how the basic ideas of fluid dynamics, discussed in Chapter 3, can be applied to rivers—how analytical idealisations can be modified to cope with the irregularities of the natural world. The initial analyses consider the broad features of flow in a river, i.e. how the velocity is related to slope, channel form and the nature of the bed, and how these features vary with discharge. It goes on to examine the habitat of bottom dwelling species, including the stability of sediment, as well as dispersion and mixing in rivers. The latter affects not only the fate of material entering a river but also the likely survival of river plankton.

5.2.1 River as a Boundary Layer Flow

A river can be viewed, in hydrodynamic terms, as a gravity driven boundary layer flow. Mixing length theory (Section 3.2.2) established the following equation for the velocity, U, near a rough surface in

Plate 1 A boulder and bedrock stream close to the source.

Plate 2 Tumbling flow in the headwaters of a mountain stream.

Plate 3 Large scale roughness flow where boulder dimensions are much greater than the water depth.

Plate 4 Shingle bars and an indication of what is implied by bankfull conditions.

Plate 5 The upper reaches of a river. This is the location of sample 3 of the particle size distribution curves shown on Fig. 3.13 (River Tweed at Lyne Ford).

Plate 6 The middle reaches of a river indicating the parabolic form of gravel rivers.

Plate 7 The lower reaches of a medium sized river. Bankfull conditions can be clearly identified and the cross-sectional form tends towards the trapezoidal when the banks are more erosion resistant.

Plate 8 Drainage ditch emphasising that the distinction between running and standing waters is not always clear cut.

terms of height above the bed, h, within a boundary layer

$$U = 5 \cdot 75 U_f \log(30h/r_p) \qquad (5.2.1)$$

where U_f is the friction velocity and r_p is a measure of the size of the roughness projections on the bed. In addition, since the mean velocity, $\bar{U}$, occurs at a height of $0 \cdot 4h$ above the bed,

$$\bar{U} = 5 \cdot 75 U_f \log(12h/r_p) \qquad (5.2.2)$$

What controls velocity in a river can be considered in terms of energy conversion—how slope, effectively potential energy per unit length, is converted into kinetic energy. The above equations for velocity in a boundary layer consider only the effect of friction due to the roughness of the bed. In a natural river, there are sources of energy loss. Leopold *et al.* (1964) identify, besides friction loss, internal distortion loss caused by boundaries that set up eddies and secondary circulations and spill resistance, associated with the occurrence of white water. Velocity in a river, therefore, is usually expressed in terms of empirical flow equations which contain a coefficient related to the various sources of energy loss. The equations below contain the parameter h_r, the hydraulic mean depth, but, when the channel is wide compared to its depth, h_r, is very nearly equal to the mean depth, $\bar{h}$ (Section 5.1.3). The most commonly used flow equations are

Chezy $\qquad\qquad \bar{U} = C\sqrt{h_r S} \qquad (5.2.3)$

Manning $\qquad\quad \bar{U} = 1/n \cdot h_r^{0.67} S^{0.5} \qquad (5.2.4)$

D'Arcy–Weisbach $\quad \bar{U}/\sqrt{gh_r S} = \sqrt{(8/f)} \qquad (5.2.5)$

Most of the older literature refers to the Chezy or Manning equations but modern practice favours the D'Arcy–Weisbach equation because it is dimensionally correct so that the value of the coefficient is independent of the units used. Because of this dimensional dependence in the other equations, care must be taken in using tabulated values of the coefficients C and n since these may refer to British units.

By balancing the gravity force acting downslope to the force resisting motion, it can be shown that the frictional velocity, U_f, is given by

$$U_f = \sqrt{gh_r S} \qquad (5.2.6)$$

Using this relation and substituting $\bar{h}$ for h_r as mentioned above, the relation between these empirical equations and the logarithmic boundary equation can be seen. For the Chezy equation,

$$C = 5 \cdot 75 U_f \log(12\bar{h}/r_p)g \qquad (5.2.7)$$

Correspondence between the analytical and empirical equations should not be taken literally but the above does imply that the Chezy coefficient has some dependence on water depth. It happens that, within the range of likely values, $\log(12h/r_p)$ is approximately equal to $h^{1/6}$. The Manning coefficient can, therefore, be expressed as

$$1/n = C\bar{h}^{1/6} \qquad (5.2.8)$$

The odd form of the D'Arcy–Weisbach coefficient, $\sqrt{8/f}$, arises because the equation was originally derived for flow in pipes. It is related to the Chezy coefficient by the following

$$\sqrt{8/f} = C/\sqrt{g} \qquad (5.2.9)$$

A great deal of engineering experience is centred on ways of estimating the coefficient n in the Manning equation. A systematic procedure for estimating n is given by Cowan (1956), viz.

$$n = (n_0 + n_1 + n_2 + n_3 + n_4)m \qquad (5.2.10)$$

where $n_0, n_1, \ldots, n_4$ express the effects of the material forming the channel, degree of channel irregularity, cross-sectional variations, obstructions and vegetation respectively. The coefficient, m, refers to the degree of meandering, varying from $1 \cdot 0$ for a straight channel to $1 \cdot 3$ for pronounced meandering. In this equation lies the engineering logic for the canalisation of rivers which has destroyed so many river habitats in lowland areas. The creation of a uniform, obstruction free channel reduces the value of n, thus increasing the velocity and discharge capacity.

More recent research has concentrated on the dependence of the coefficient in the D'Arcy–Weisbach equation to the relative roughness, i.e. the ratio of the water depth to the bed particle size, usually expressed as the d_{84} size (see Section 3.3.3). In terms of the theoretical boundary layer equation, the D'Arcy–Weisbach equation can be written as

$$\sqrt{8/f} = \bar{U}/\sqrt{g\bar{h}S} = 5 \cdot 75 \log(12h/r_p) \qquad (5.2.11)$$

Comparison of a number of investigations on the effect of relative roughness on flow in rivers indicates that one of the earliest, that due to Wolman (1955), is representative and that, although the original equation was based on h/r_p values no greater than 15, it can be extrapolated up to a h/r_p value of 100. Wolman's equation, in the form used here, is

$$\bar{U}/\sqrt{g\bar{h}S} = 5.66 \log(\bar{h}/r_p) + 2.83 \qquad (5.2.12)$$

This, together with the equivalent values for theoretical roughness given by eqn (5.2.11), is plotted in Fig. 5.8. Also plotted is the energy loss ratio which is simply the ratio of the measured to theoretical flow coefficients. This illustrates the earlier comments about sources of energy loss in rivers. Where the bed particles are small compared to the water depth, the ratio is virtually constant and friction accounts for three quarters of the total losses although other sources of energy loss, such as extensive stands of vegetation, may alter this. Where there are stones on the bed equal in size or even greater than the water depth, friction losses become less important than the internal distortion losses and the occurrence of white water.

Equation (5.2.12) establishes that the slope, water depth and size of the material forming the bed are what determine the velocity of the flow. The value of $\bar{h}/d_{84}$ equal to 4 can be used to separate flow types in rivers. For $\bar{h}/d_{84} > 4$, motion in rivers can be considered as described in Section 3.2. Mixing length theory, for example, can be assumed to apply and the conventional pattern of velocity distribution can be observed (Fig. 5.9). Despite the fact that the D'Arcy–Weisbach equation is still valid, motion at lower values of $\bar{h}/d_{84}$ is obviously different. The water surface is no longer smooth, white water may occur and the velocity distribution across the width of a river is determined by the more or less random pattern of stones on the bed.

Peterson and Mohanty (1960) distinguish three forms of large scale roughness flow when $\bar{h}/d_{84} > 4$, viz.

—tranquil flow where the surface is uneven, no white water occurs and the flow is characterised by the shimmering in bright sunlight;
—tumbling flow where there is an alternating pattern of shooting flow over the surface of stones and breaking waves with the occurrence of white water;
—rapid flow occurring throughout, the pools between stones being swept away.

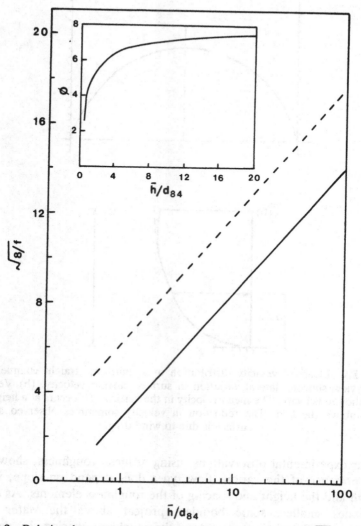

Fig. 5.8 Relation between the D'Arcy-Weisbach coefficient, $\sqrt{8/f}$, and the relative roughness, $\bar{h}/d_{84}$. The full line is the observed relation in rivers (eqn (5.2.12)) and the dashed line is the theoretical relation in a boundary layer flow (eqn (5.2.11)). The insert shows the ratio, ϕ, of the observed to theoretical coefficients.

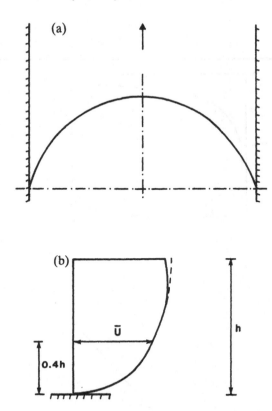

Fig. 5.9 Idealised velocity distribution in a uniform, straight channel. (a) Plan view showing lateral variation in surface current velocity. (b) Vertical variation in velocity. The mean velocity in the vertical, $\bar{U}$, occurs at a height of $0\cdot4h$ above the bed. The reduction in velocity sometimes observed at the surface is due to wind drag.

Their experimental observations, using artificial roughness, show that the presence of the various flow types does depend on slope, water depth and the height and spacing of the roughness elements. As $\bar{h}/d_{84}$ becomes smaller, some boulders project above the water line. Bathurst (1978) gives a measure of the roughness spacing, λ_*, defined as the ratio of the basal area of projecting boulders at the water line to the total bed area, where

$$\lambda_* = 0\cdot36 \log(1\cdot52 d_{84}/\bar{h}) \qquad (5.2.13)$$

Note that the relative roughness term is inverted compared to normal practice.

5.2.2 Stage Discharge Curves

A stage discharge curve is the relation between the water level in a river and the flow rate or discharge. Such a curve provides the link between the climate and catchment characteristics, expressed as discharge, and the nature of the habitat. If the discharge and water level are known at a particular section, then ecologically more relevant factors, such as velocity or bed stress, can be calculated.

Most observed stage discharge curves are constructed for the opposite purpose of measuring streamflow, i.e. the level is measured and the observed curve provides a rating or calibration by which records of water level are converted to discharge. Such curves may not be typical of a river. The main concern is precision and the river length for which the curve is derived is under what is termed section control—it is upstream of a bar or other obstacle so that there is a large change in water level for a given change in flow. To understand the nature of a river habitat, the stage discharge curve should be applicable to as great a length of river as possible, the level–flow relation determined by the general features of the channel and the previous flow equations applicable. Lower standards of accuracy are acceptable.

Most observed relations between the discharge, Q, and the gauge reading, h', are of the form

$$Q = a(h' - c)^b \qquad (5.2.14)$$

see Fig. 5.10. The correction factor, c, has to be included since a reading of zero on the gauge does not usually coincide with zero flow in the river. Where the object is to establish an analytical relation between level and flow, the correction factor can be omitted and the gauge reading replaced by the height above the bed, h, so that

$$Q = ah^b \qquad (5.2.15)$$

Rough estimates of the exponent are easily obtained. Discharge is the product of the mean velocity, $\bar{U}$, and the cross-sectional area, A_h. Applying Manning's eqn (5.2.4), where the channel is wide compared to its depth so that the hydraulic mean depth, h_r, is equal to the depth, h, $\bar{U}$ is proportional to $h^{0.67}$. The relation between A_h and h depends on the channel form. In a parabolic channel, for example, $A_h = 2hW/3$ and $W = W_{bx}\sqrt{h/h_{bx}}$ (Table 5.3), so that A_h is proportional to $h^{1.5}$. The discharge, therefore, is proportional to $h^{0.67}$ times $h^{1.5}$, i.e. the

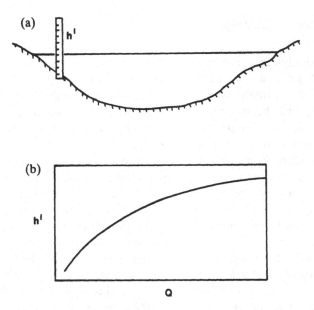

Fig. 5.10 General form of stage-discharge curves, relating the staff gauge reading, h', to the discharge, Q.

exponent, b, is equal to 2·17. In principle, the coefficient, a, can also be derived if it is assumed that the Manning coefficient and the slope remain constant over the range of flows. This is not always the case and direct measurement is preferable.

When the discharge exceeds the bankfull value, the flow is no longer confined within the river channel and there is a greater increase in discharge for the same increase in water level. This is seen more clearly when the curve is plotted with logarithmic scales (see Fig. 5.11).

5.2.3 Dissolved and Suspended Loads

The concentrations of dissolved and suspended particulate material in river flow are often expressed as power functions of the discharge, i.e. $C = aQ^b$. It must be stressed that such relationships obscure a number of processes occurring on the catchment and within the river itself which are not discussed here. Measured quantities of dissolved and particulate matter include both the natural products of weathering and

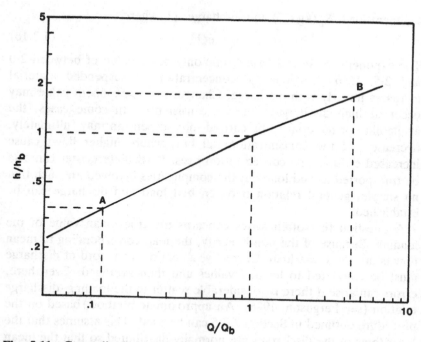

Fig. 5.11 Generalised stage discharge curve for 13 stations in the United States in which both the level, h, and the discharge, Q, are expressed as proportions of bankfull values (based on Leopold *et al.*, 1964). A refers to the mean flow and B to a flood with a return period of 50 years. For flows less than bankfull, the slope of the line corresponds to an exponent value of 2·25.

erosion and pollutants. The instantaneous material discharge rate or load (kg s^{-1} or t day^{-1}) is the product of concentration (kg m^{-3}) times discharge (m^3 s^{-1}).

In a world wide review of the total dissolved solids concentration, Walling and Webb (1983) show that the average value of the exponent, b, is −0·17 with very few values being less than −0·60. In general, the concentration of dissolved solids decreases with discharge. The loading, therefore, is, on average, proportional to $Q^{0.83}$. The exponent in the concentration discharge relation for particular elements can vary widely and may be positive, concentration increasing with discharge. See, for example, Edwards (1973) and Smith (1976).

The quantity of suspended particulate matter carried by a river is usually measured by a sediment rating curve, i.e. the relation between sediment load, S_q (kg s^{-1}) and discharge, Q, where

$$S_q = aQ^b \qquad (5.2.16)$$

The exponent, b, in this case, commonly has a value of between 2·0 and 2·5. If $b = 2$, then the concentration of suspended material increases linearly with discharge. Changes in the exponent value may occur at high discharges but not consistently. In some cases, the availability of material to be carried into suspension may, ultimately, decrease and the concentration fall but, where higher flows cause increased erosion, the concentration rises. Particulate matter can also be transported as bed load but the complexities involved are such that no simple, general relation between bed load and discharge can be established.

A question that often arises concerns the true mean value of the loading. Because of the non-linearity, the load corresponding to mean flow is not the mean load. Correctly, a continuous record of discharge must be converted to loading values and then averaged. Even here, errors can arise if there is considerable scatter in the loading–discharge relation (see Ferguson, 1986). An approximate method, based on the procedure outlined in Section 2.2.5 can be used. This assumes that the logarithms of the discharges are normally distributed so that the mean and standard deviation of a related variable can be estimated. The procedure is identical to that for velocity which is given in more detail in the numerical example below.

5.2.4 Numerical Example

A number of river features discussed earlier are illustrated in this example. Conditions at a particular point in a river are examined.

1 Initial data
Bankfull discharge, Q_b 500 m^3 s^{-1}
Average discharge, $\bar{Q}$ 50 m^3 s^{-1}
Catchment area, A_c 2500 km^2
Channel slope, S 1·0 m km^{-1}
The cross-sectional form is parabolic and the variability index is 30.

2 Channel dimensions
—width at bankfull:

$$W_{bx} = 3·5\sqrt{Q_b} = 78·26 \text{ m} \qquad (5.1.7)$$

—sectional area at bankfull:

$$A_b = [1 \cdot 73 Q_b / S^{0 \cdot 28}]^{0 \cdot 826} = 266 \cdot 67 \ m^2 \qquad (5.1.8)$$

—maximum depth at bankfull:

$$h_b = 3A_b / 2W_{bx} = 5 \cdot 11 \ m \qquad (5.1.9a)$$

To plot the parabolic section, the height of the bed above its lowest point, y, at a distance, x, from the centre line is given by $y = ax^2$ where $a = 4h_b / W_{bx}^2 = 0 \cdot 003 \ 34$ (see Fig. 5.12).

3 Stage–discharge curve

From eqn (5.2.15), $Q = ah^b$, and, for a parabolic section, $b = 2 \cdot 17$. Applying conditions at bankfull,

$$a = 500 / (5 \cdot 11)^{2 \cdot 17} = 14 \cdot 51$$

—depth at average flow: $50 = 14 \cdot 51 h^{2 \cdot 17}$, i.e. $h = 1 \cdot 79 \ m$

Relative depth at mean flow $= 1 \cdot 79 / 5 \cdot 11 = 0 \cdot 35$. This is virtually identical to the ratio indicated on Fig. 5.11.

4 Conditions at other water levels

—dimensions: top width, $W = W_{bx} \sqrt{h/h_b} = 34 \cdot 62h$ (Table 5.3), sectional area, $A = 2hW/3$

—velocity: given the discharge from the stage–discharge curve, the mean velocity, $\bar{U} = Q/A$. The values of the variables at other water levels are listed in Table 5.6.

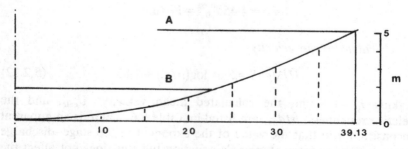

Fig. 5.12 Computed half section of a parabolic channel. A represents bankfull conditions and B average flow.

Table 5.6   River characteristics at a particular section

h_b (m)	Q ($m^3 s^{-1}$)	W (m)	A (m^2)	$\bar{U}$ ($m s^{-1}$)	$\bar{h}/d_{84}$	U_{calc} ($m s^{-1}$)	θ_m	θ
0·5	3·22	24·48	8·16	0·39	2·80	0·30	0·007	0·005
1·0	14·51	34·62	13·09	0·63	5·60	0·56	0·015	0·010
1·5	34·98	42·40	42·42	0·82	8·41	0·80	0·023	0·015
2·0	65·30	48·96	65·31	1·00	11·21	1·00	0·030	0·020
2·5	105·98	54·74	91·28	1·16	14·01	1·19	0·038	0·025
3·0	157·41	59·96	119·98	1·31	16·81	1·37	0·045[a]	0·030
3·5	219·93	64·77	151·21	1·45	19·62	1·53	0·053	0·035
4·0	293·86	69·24	184·73	1·59	22·42	1·70	0·061	0·040
4·5	379·44	73·44	220·43	1·72	25·20	1·84	0·068	0·045[a]
5·11	499·97	78·26	266·74	1·87	28·64	2·03	0·077	0·051

[a] Approximate critical value for the onset of movement on the bed.

—hydraulic geometry: The power laws relating variables to discharge are as follows

$$h_b = 0·291Q^{0·46}$$

$$W = 18·69Q^{0·23}$$

$$\bar{U} = 0·273Q^{0·31}$$

The sum of the exponents is exactly 1. The mean depth $\bar{h} = 2/3h_b$ and the product of the coefficients $\times 2/3 = 0·99$.

5   Bed particle size

For $A_c > 1000\,km^2$, $d_{50}/S = 40$ approximately (Section 5.1.4) so that $d_{50} = 40\,mm$. From eqn (3.3.1),

$$d_{84} = 4·45d_{50}^{0·89} = 119\,mm$$

6   Calculated mean velocity

$$\bar{U}/\sqrt{ghS} = 5·66 \log(h/r_p) + 2·83 \qquad (5.2.12)$$

Taking $r_p = 119\,mm$, the calculated mean velocity, U_{calc}, and the relative roughness, $\bar{h}/d_{84}$, are listed in Table 5.6. There is an apparent inconsistency in that the value of the exponent in the stage–discharge curve is derived from Manning's equation but this does not affect the validity of the calculated velocities.

7 Entrainment function

$$\theta = \tau/[(\rho_s - \rho)gd_{50}] = \rho ghS/(\rho_s - \rho)gd_{50}$$
$$= 0 \cdot 001h/(1 \cdot 65 \times 0 \cdot 04) = 0 \cdot 0151h \tag{3.4.1}$$

Two values of θ are listed in Table 5.6. θ_m refers to the centre of the river $(h = h_b)$ and θ to the mean over the bed $(h = \bar{h})$. Some movement of the bed occurs in the middle of the river when the discharge is about three times the average flow. Movement over most of the bed occurs when the flow is somewhat less than bankfull.

8 Properties of the log transformed discharges
—standard deviation:

$$\sigma_q = 0 \cdot 289 \log v_i - 0 \cdot 027 = 0 \cdot 40 \tag{2.2.2}$$

—Quenouille's correction:

$$\bar{L}_q + 1 \cdot 15\sigma_q^2 = \log \bar{Q} \tag{2.2.3}$$

i.e.

$$\bar{L}_q + 1 \cdot 15(0 \cdot 40)^2 = \log 50, \qquad \bar{L}_q = 1 \cdot 515$$

The equivalent flow, i.e. the flow exceeded 50 per cent of the time, is $32 \cdot 73 \text{ m}^3 \text{ s}^{-1}$.

—flow duration curve:
Using the properties of the normal distribution that the flow exceeded 84 per cent of the time is 1 standard deviation less than the mean and so on, the following are obtained

	$\log Q$	Q $(\text{m}^3 \text{s}^{-1})$
Q_{98}	0·715	5·19
Q_{84}	1·115	13·03
Q_{16}	1·915	82·2
Q_2	2·315	206·5

see Fig. 5.13. The arithmetic mean flow $(50 \text{ m}^3 \text{ s}^{-1})$ is exceeded 32 per cent of the time but the variability index recovered $= Q_1/Q_{90} = 270/10 \cdot 3 = 26 \cdot 2$. The error is believed to be due to the fact that eqn (2.2.2) is an empirical one derived from actual flow duration curves and is not based on the inherent properties of the normal distribution. The method of plotting exaggerates any error in standard deviation since it involves pivoting about the 50 per cent point.

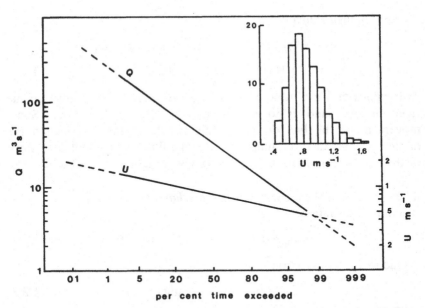

Fig. 5.13 Computed duration curves for discharge, Q, and velocity, U. Inset shows the frequency distribution of velocities to natural scales.

9 Properties of variables related to discharge

For illustration, mean velocity is examined, i.e. $\bar{U} = 0.273Q^{0.31}$. For a variable related to Q by $Y = aQ^b$, the mean of the log transformed variable is given by

$$\bar{L}_y = b\bar{L}_q + \log a \qquad (2.2.5a)$$

i.e. for velocity,

$$\bar{L}_y = 0.31 \times 1.515 + \log(0.273) = -0.094$$

Similarly, for the standard deviation,

$$\sigma_y^2 = b^2\sigma_q^2 = (0.31)^2(0.40)^2 = 0.0154 \qquad (2.2.5b)$$

i.e.

$$\sigma_y = 0.124$$

Velocity duration curve:
Using the standard properties of the normal distribution as before, the

following values are obtained

	$\log \bar{U}$	$\bar{U}$ (m s^{-1})
U_{98}	-0.342	0.45
U_{84}	-0.218	0.61
U_{50}	-0.094	0.81
U_{16}	0.030	1.07
U_2	0.154	1.42

The flow duration curve is plotted in Fig. 5.13. Also shown is the frequency distribution in natural units, derived graphically from the duration curve.

The arithmetic mean velocity can be obtained using Quenouille's correction, i.e.

$$\log(\bar{U}) = \bar{L}_y + 1.15\sigma_y^2 = -0.094 + 1.15(0.124)^2 = -0.763$$

This corresponding to a velocity of 0.84 m s^{-1}. The velocity associated with the mean discharge $= 0.273(50)^{0.31} = 0.92$ m s^{-1}.

5.3 HYDRODYNAMIC FEATURES OF RIVER HABITATS

5.3.1 Introduction

This section considers how the basic concepts of fluid motion can be used to elucidate the nature of river habitats. There are not always direct links to ecological observations but rather the intention is to show how hydrodynamics provide the starting point for systematic habitat description. Sediment stability is all important. The system of river classification in Section 5.1.2 stresses activity as a distinguishing feature and the traditional division of rivers into eroding, transporting and depositing reaches, based on qualitative observation, is an oversimplification. An analytical approach that takes flow variability into account is desirable. Related to sediment stability is the downstream flow of organic particulate matter—the same physical principles applying to both organic and inorganic particles.

Also considered in this section is what may be termed micro-hydraulics by analogy with micro-meteorology in terrestrial environments. This examines the detailed flow pattern round bed particles forming the habitat of benthic invertebrates. The third hydrodynamic feature concerns dispersion and mixing in rivers, e.g. the validity of the advection–diffusion equation and obtaining the residence time distribution for a river reach.

5.3.2 River Stability

Distinguishing two forms of sediment movement, viz. bed load where the particles are supported by inter-particle contact and suspended load where particles are supported by hydrodynamic forces, the sediment stability question can be expressed formally in terms of two mass balance equations. Within a reach, the change in amount with time, dM/dt, equals the gains less the losses.

For the bed,

$dM_b/dt =$ bed load in + material deposited − material carried into suspension − bed load out

For the suspended material,

$dM_s/dt =$ suspended load in + material carried into suspension − material deposited − suspended load out

There is no question of solving these equations directly. They are helpful, however, in clarifying what is happening. Much of the confusion about the interpretation of river stability relates to the different time scales over which the equations are integrated. Ecologically, the short term is important. If there is short term deposition, fish eggs, for example, may be smothered and deprived of oxygen. Another view of equilibrium, much used by engineers, occurs when rivers and canals are said to be in regime (Lacey, 1929). Over a seasonal cycle, many different states of sediment movement may occur. A river is in regime if, over the seasonal cycle, there is no net erosion or accumulation—no maintenance work is required. It is possible, therefore, for a river to be stable over the longer, geomorphological time scale and yet unstable in terms of habitat.

Further insight into river stability can be obtained by linking the mass balance equations to critical values of Shield's entrainment function, θ (see Section 3.4.4). It is stressed again that the exact critical values of θ are still the subject of considerable discussion and dependent on local features (see, for example, Carling, 1983). The object is to establish a basis for identifying the different stability regimes that can occur. Since $\theta = \tau/(\rho_s - \rho)gd$ and $\tau = \rho ghS$, we can write

$$\theta = h/dS\rho/(\rho_s - \rho) \qquad (5.3.1)$$

Critical values of θ, therefore, can be displayed on a plot of h/d_{50} against S. Figure 5.14 shows the critical values for coarse material ($d_{50} > 2$ mm, approximately) which are independent of particle size. The top right hand corner of the figure relates to catastrophic

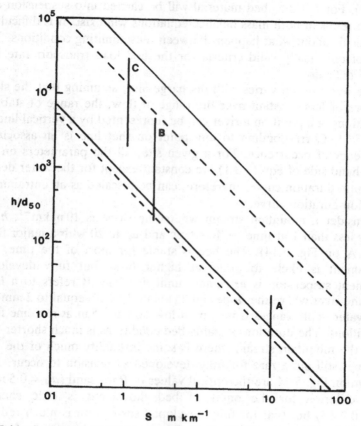

Fig. 5.14 Sediment stability in rivers based on the relation between relative roughness, h/d_{50}, and channel slope, S. The full lines represent the critical values of the entrainment function, θ, for the onset of bed movement and fully developed suspension for coarse material ($d_{50} > 2$ mm), i.e., 0.045 and 2.5 respectively. The dashed lines are the equivalent for sand ($d_{50} = 0.5$ mm), i.e., 0.0335 and 0.83. The vertical lines represent the range of conditions in typical rivers (see text for details).

conditions such as occur following the bursting of an ice dammed lake. The bottom left hand corner refers to ultra stable conditions—moss covered stones whose presence is not the direct result of fluvial processes at all.

For $\theta < 0.045$, there is no bed load movement and the only suspended matter is that introduced from outside the reach in question. There is no restriction on the size of this material and, since resuspension is unlikely, there will be deposition as given by eqn (3.4.5). For $\theta > 2.5$, bed material will be carried into suspension and all the terms in both mass balance equations will exist. It is difficult to be specific about what happens between these limiting conditions. The lack of universally valid criteria for the bed load transport rate is a major difficulty.

The water depth varies with discharge and, assuming that the slope is more or less constant over the range of flow, the range of stability conditions at a point on a river can be represented by a vertical line on Fig. 5.14. Corresponding to any point on that line is an associated frequency of occurrence. For a given site, all the parameters on the right hand side of eqn (5.3.1) are constant except for the water depth. A depth–duration curve, therefore, can be rescaled as an entrainment function duration curve.

Consider a mountain stream where the slope is 10 m km^{-1}, h/d_{50} being less than 1 at times of low flow and up to 20 with a major flood (line A on Fig. 5.14). The bed is stable for most of the time, bed movement is likely to occur at higher flows but fully developed sediment suspension is extremely unlikely. Line B refers to a large lowland river where the slope is 0.25 m km^{-1}, d_{50} is equal to 2 mm and the water depth ranges from 1 m at low flow to 15 m at extreme flood conditions. The duration of stable bed conditions is much shorter than with the mountain stream, there is some instability much of the time yet it is still quite rare for fully developed suspension to occur. Also shown on Fig. 5.14 are the critical values of θ for sand ($d_{50} = 0.5$ mm). The criterion for the onset of bed movement is little changed ($\theta = 0.0395$) but that for fully developed suspension is much reduced ($\theta = 0.83$). Line C shows that, for a sand bed with a slope of 0.1 m km^{-1} and the same water level range as in the previous example, there is some movement all the time and that full developed suspension is relatively common.

A vertical line, representing the range of discharges occurring, is not a precise measure of the stability characteristics of a reach but it is a useful indicator, particularly for between site comparisons. By plotting successive vertical lines corresponding to a series of locations along the length of a river, it is, in principle, possible to generate an alternative to the qualitative division into eroding, transporting and depositing regimes. Such a scheme, a sediment stability frequency diagram, is simply a plot of the frequency of occurrence of different modes of sediment movement against distance from the source and is shown diagrammatically in Fig. 5.15. Such an approach emphasises that many rivers are incomplete in that the full range of conventional features does not occur. For example, Britain's largest river in terms of discharge, the River Tay, still has an upland character, dominated by bed load, when it enters its estuary (Maitland and Smith, 1987).

The stability regimes defined by Fig. 5.14 also emphasise the erratic nature of the movement of sediment through a river system. The amount of sediment delivered at a point depends not only on these transport mechanisms but also on the availability of material to be transported. Much of the suspended sediment, particularly in the upper reaches, may be derived from specific events such as landslips and bank erosion. Simple relations such as those linking suspended solids concentration to discharge are a gross oversimplification.

Since, even for inorganic sediments, the critical values of the entrainment function are uncertain, the likelihood of establishing exact values for the varied shapes of organic particles is small. There is a

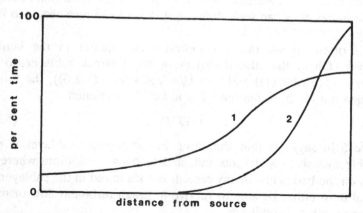

Fig. 5.15 Diagrammatic representation of sediment stability–frequency curves showing how the percentage of time different modes of sediment movement occur along the length of a river.

limit to the value of hydrodynamic analysis. It seems preferable to view the entrainment function for inorganic sediments as a measure of hydrodynamic state. If observations of stream drift, for example, could be related to this value of the entrainment function, then a rational, yet feasible, analysis may be possible.

5.3.3 Micro-hydraulics

Micro-hydraulics are concerned with what happens in the zone of direct interaction between flowing water and the particles forming the river bed—so called because of the similarity to micro-meteorology in terrestrial environments. It is the habitat of benthic invertebrates, the root zone of higher plants and the site for fish spawning beds. Once the critical value of the entrainment function is exceeded, disturbance is the dominant feature so that the concern is with stable bed conditions. This section is simply a sketch of the possible basis for micro-hydraulics—establishing a link between conventional hydrodynamics and animal behaviour.

From the animal's point of view, two features appear to be of paramount importance. The first is survival—ensuring that it is not swept away. The second is access to food supply, much of which is drifting past in the current. Ultimately, the criterion for survival for an animal is the same as that for the stability of inorganic sediments, i.e. the resisting force must be greater than the disturbing force. Most animals have a behavioural option so that, if the conditions become unfavourable, they can seek shelter but this may reduce access to food supply.

It is often argued that invertebrates can shelter in the laminar sublayer. Given that the thickness of the laminar sublayer, δ', $= 11 \cdot 5 v/U_f$ (eqn (3.2.12)) and that $U_f = \sqrt{ghS}$ (eqn (5.2.6)), the following equation for δ' in mm and S in m km^{-1} is obtained

$$\delta' = 0 \cdot 133/\sqrt{hS} \tag{5.3.2}$$

Figure 5.16 suggests that sheltering in the laminar sublayer is only possible with shallow streams and/or flat slopes—conditions where the stress on the bed is low in any case. If not sheltered in the sublayer, an animal must either be able to survive in rough turbulent flow or move to more sheltered conditions.

The ability to find shelter depends on the size and layout of particles on the bed. With uniform, fine sediment, the only shelter available is

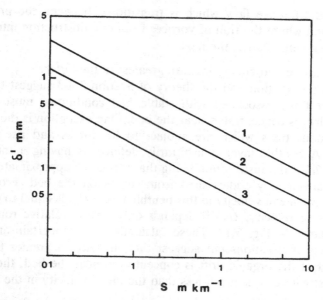

Fig. 5.16 Relation between the thickness of the laminar sub-layer, δ', and the channel slope, S, for different water depths. Line 1, 0·25 m; line 2, 1·0 m; and line 3, 5 m.

by burrowing. Where the sediment particles are much larger than the animal size, sheltering is possible as a result of the flow pattern around the stones. The available shelter depends not only on the stone size but also on their proximity to each other and the relative roughness, h/d_{84}. Section 5.2.1 indicates that the nature of the flow changes when $h/d_{84} = 4$, approximately, and distinguishes different types of large scale roughness when $h/d_{84} < 4$. For larger values, Morriss' (1955) classification of roughness types provides a starting point for estimating the available shelter. He considers both the size and spacing of the roughness elements which determine the nature of the eddies formed in their lee, distinguishing:

—isolated roughness flow where the trail of vortices behind one stone is completely dissipated above the next one;
—quasi-smooth flow where the stones are sufficiently close together that the flow skims over their crests forming a pseudo wall, the water between being virtually static;

—wake interference flow which is transitional between the previous point and where the trail of vortices from one obstruction interferes with that generated at the next.

The turbulence and energy loss are greatest in the latter.

Casual observation and the theory of Section 3.4.3 suggest that, at the lower flows associated with stable bed conditions, most of the organic drift is concentrated near the bed. One exception is the leaves of deciduous trees which are subject to lift forces and rise to the surface. A benthic layer is arbitrarily defined as having a thickness equal to $2d_{84}$, the justification being that this is the approximate depth directly affected by eddies shed from stones on the bed. From eqn (5.2.12), the mean velocity in this benthic layer, $\bar{U}_B$, is equal to $4 \cdot 53 U_f$. The relative velocity, $\bar{U}_B/\bar{U}$, depends only on the relative roughness and is shown in Fig. 5.17. These calculations are uncertain and lack supporting observations but suggest that, if h/d_{84} is greater than 10 and most of the organic drift is concentrated near the bed, the speed of downstream drift is much less than the mean velocity in the river.

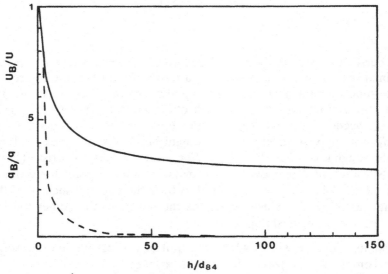

Fig. 5.17 Relative velocity, $\bar{U}_B/\bar{U}$ and discharge, $\bar{q}_B/\bar{q}$, in the benthic layer. The full line is the relative velocity and the dashed line is the relative discharge.

5.3.4 Dispersion and Mixing in Rivers

When a quantity of tracer is suddenly injected into a river, the observed concentration–time curve at a point downstream does not have symmetrical, normally distributed form implied by the simple advection–diffusion equation (see Section 3.2.4). The peak concentration may be higher than predicted and there is a much longer tail after the passage of the peak (see Fig. 5.18). This is usually explained by the existence of a dead zone, i.e. a river reach is assumed to be divided into a core or main stream volume where the standard advection–diffusion processes apply and the dead zone where injected tracer is trapped in pools, backwaters and behind large rocks. The reduced main stream volume explains the more rapid peak and the tail is caused by the slow release of tracer from storage.

Formally, eqn (3.2.33) only applies to the main stream volume and has to be modified by including a term describing transfer between the main stream and the dead zone. A second equation is required to account for the concentration change within the dead zone (see, for example, Thackston and Schnelle, 1970). Such a procedure is not very satisfactory. The mathematics are complex and, if reasonable accuracy

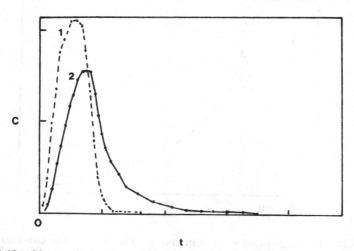

Fig. 5.18 Observed concentration time curves at two points on a river below a sudden injection of tracer (based on Young and Wallis, 1987). Curve 2 refers to the further downstream point.

is required, on-site observations are necessary, longitudinal dispersion coefficients being difficult to predict.

Young and Wallis (1986) propose a much simpler alternative in what they term the aggregated dead zone model. The basis of this is that the response in a river reach to a sudden injection of tracer can be considered in two parts, viz.

—the response as a result of the mean motion (advection) can be represented by a simple lag, an interval of time between injection and the occurrence of the peak;
—the effects of dispersion and mixing are equivalent to assuming instantaneous mixing within the dead zone (see Fig. 5.19).

The form of the concentration–time curve between injection and occurrence of the peak is not specified. The observations of Young and Wallis indicate that, for a given reach, both the lag and the dead zone volume can be accurately predicted from the discharge. There is a further indication, not yet fully substantiated, that the dead zone volume as a fraction of the total is more or less constant (0·30, approximately). As might be expected, the ratio in artificial channels is somewhat lower.

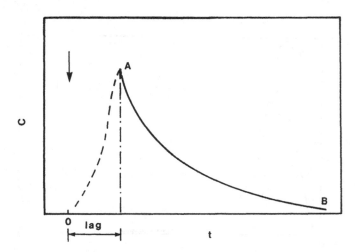

Fig. 5.19 The aggregated dead zone model. The form of the concentration–time curve can be predicted if the lag and the volume of the dead zone are known. The curve AB represents instantaneous mixing within the dead zone volume.

The form of Fig. 5.8 could probably be reproduced by assuming the reach to consist of a number of basins in series (see Section 4.3.3) but there is no physical logic for this. Knowing the response to a sudden injection of tracer, it is a simple matter to calculate the residence time distribution (Section 4.2.2) and thus the effect of dispersion and mixing on internal reactions. The duration of exposure to different concentrations of toxins can also be estimated.

6

Hydraulic Characteristics of Lakes

6.1 LAKE MORPHOLOGY

6.1.1 Introduction

The origin and morphology of lakes is discussed, in detail, by Hutchinson (1957) and the primary concern here is only with the features of lake size and shape that directly relate to water movement. Compared to rivers whose dimensions are, normally, related directly to present day flows, most lakes have been created by past geological events. There is, therefore, no logical basis for any correlation between lake dimensions and present day conditions. Statistical relations between the lake dimensions themselves have been described by a number of investigators. Gorham (1958), for example, by making the simple distinction between rock and drift basins, found relations between surface area and mean depth and between maximum and mean depth in the Scottish lochs surveyed by Murray and Pullar (1910).

6.1.2 Standard Measures

Largely following Hutchinson's (1957) discussion, a number of measures of lake size and form have become accepted as standard. Figure 6.1 shows the bathymetry of two Scottish lochs—Ness and Leven—which typify Gorham's separation of rock and drift basins and Table 6.1 shows the standard measures for each. The volume is determined by adding the volumes of the slices between underwater contours and the mean depth is defined as the total volume divided by the surface area. The shoreline development is the ratio of the

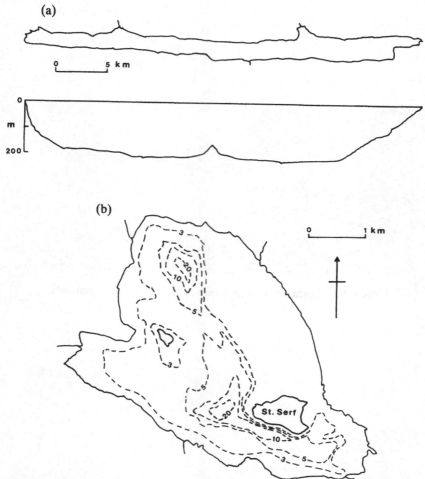

Fig. 6.1 The bathymetry of two Scottish lochs. (a) Map and longitudinal section of Loch Ness; (b) bathymetric map of Loch Leven, showing underwater contours in metres.

shoreline length to the circumference of a circle having the same area as that of the lake.

Another way of summarising bathymetric data is to plot area and volume against depth below the surface (Fig. 6.2). In shallow lakes where the range of water level fluctuation is large compared to the total depth, it is important that the relation between the survey datum and the modal water level is clearly defined. A theoretical analysis of

Plate 9 A typical rock basin (the north end of Loch Lomond).

Plate 10 A shallow drift basin (Loch Laidon).

Plate 11 An upland cirque or corrie lake (Loch Brandy).

Plate 12 A lowland lake (Burnham Mere, Holme Fen National Nature Reserve).

Table 6.1   Standard measures of lake morphology

		Ness[a]	Leven[b]
Surface area	km^2	56·4	13·3
Mean depth	m	132·0	3·9
Maximum depth	m	229·8	25·5
Volume	m$^3 \times 10^6$	7 450·0	52·4
Length	km	39·0	5·9
Breadth	km	1·5	2·3
Shoreline length	km	86·0	18·5
Shoreline development	—	3·23	1·43

[a] Smith *et al.* (1981).
[b] Smith (1974).

the form of such curves is given by Junge (in Hrbacek, 1966) who shows that the curves can be generalised in terms of the relative depth, z/z_{max}, and a single parameter that distinguishes the various geometric forms of a lake—cones, hyperboloids, paraboloids and ellipsoids.

6.1.3   Hydrodynamic Measures

The standard presentation of lake morphology does not provide all the information relevant to the influence of water movement. Reference has already been made to water level fluctuations and, since area–depth curves are virtually linear within the normal range of level fluctuation in natural lakes, the area change per unit change in water level is a useful measure. For Lochs Ness and Leven, the values are 0·25 and 2·82 km^2 m^{-1} respectively. The difference between the two is further emphasised when the effects are expressed as percentage change in surface area per metre change in level—0·45 and 21·2 per cent respectively. Similar ratios for volume change can, obviously, be calculated.

The form of a lake affects its susceptibility to wave action and, thus, the resultant sediment stability. This is discussed further in Section 6.8.3 where it is shown that the dynamic ratio, the square root of the lake area divided by the mean depth, proposed by Hakanson (1982) can be used to distinguish the sediment regimes in a lake. The greatest difficulty in finding morphometric measures that express the influence of water movement concerns the shape of the lake surface. Hutchinson

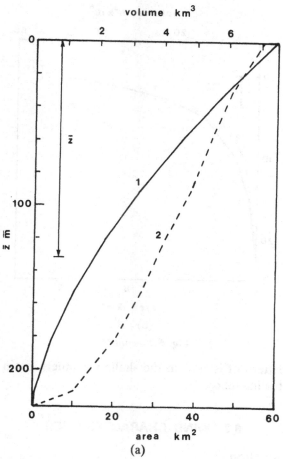

Fig. 6.2 Volume and area-depth curves. (a) Loch Ness; (b) Loch Leven, the vertical line at A indicating the range of level fluctuation. In each case, line 1 represents volume and line 2 area.

(1957) refers to the lake articulation, the ratio of the area of bays and inlets to the total lake area, but dismisses it as little used nowadays. Ultimately, the problem is not that of morphology alone but the combined effects of morphology and hydrodynamics so that simple measures may not be feasible. Comparing Loch Ness (Fig. 6.1*a*) and Loch Lomond (Fig. 6.3), it is clear that the influence of lake form on the water motion in Loch Ness is small—it can be considered as a single basin—while Lomond is, effectively, three linked basins and

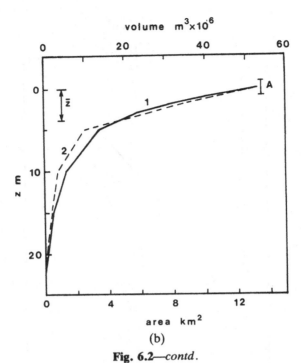

Fig. 6.2—*contd*.

where the pattern of islands in the shallow southern basin must have an appreciable influence.

6.2 WIND CHARACTERISTICS

6.2.1 Introduction

Wind is the major cause of motion in lakes but conventional analyses of wind data are not always appropriate for limnological use. What is required is an analysis of the wind regime over a lake, analogous to that for stream flow, that can be used to predict the nature and frequency of occurrence of different forms of lake motion. Even if continuous records of wind speed over water are available, long records are necessary to establish general patterns. Only winds in the UK are considered here, primarily to indicate the types of analyses that may be useful. Virtually all available wind data refer to sites on land and it is assumed that generalisations derived for land stations can be applied over lakes with appropriate corrections.

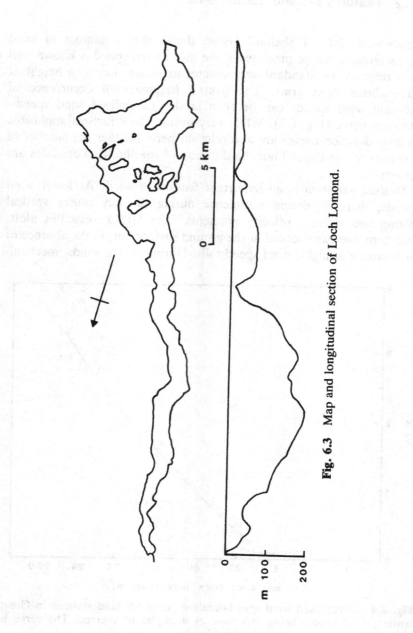

Fig. 6.3 Map and longitudinal section of Loch Lomond.

6.2.2 Features of Land Station Data

Analysis of data in Shellard (1968) shows that a number of wind characteristics can be predicted if the mean wind speed is known. All data refer to the standard anemometer exposure, i.e. to a height of 10 m above short grass. The overall frequency of occurrence of different wind speeds can be seen in the generalised wind speed–duration curve (Fig. 6.4). While very useful, the criticisms applicable to flow–duration curves are also relevant here. Neither the number of reversals of wind speed nor the duration of specific wind episodes are known.

Diurnal variation is an important feature of wind. At lower wind speeds, thermally-driven turbulence during the day causes vertical mixing and reduces velocity gradients. The higher velocities aloft, therefore, are experienced at the ground surface but, in the absence of such mixing at night, wind speed falls. During strong winds, mechani-

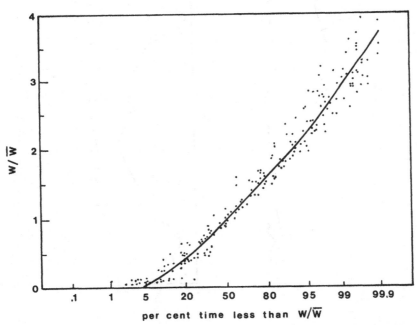

Fig. 6.4 Generalised wind speed-duration curve for land stations in Great Britain, wind speeds being expressed as multiples of average. The curve is fitted by eye.

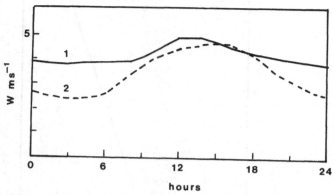

Fig. 6.5 Diurnal variation in wind speed at Kew (based on Chandler and Gregory, 1976). Line 1 represents average winter conditions (Dec.–Feb.) and line 2 average summer conditions (Jun.–Aug.).

cal turbulence is dominant and much of the daily variation is suppressed. The influence of solar heating on diurnal wind patterns can be seen in the comparison of winter and summer averages (Fig. 6.5). In tropical climates, the daily wind pattern can be so pronounced that diurnal lake mixing features can occur (see Viner and Smith, 1973).

The general windiness of a particular location can be measured as the duration of gales (wind speeds > 17 m s^{-1}). This can be predicted from mean wind speed (Fig. 6.6a), the seasonal pattern being reasonably consistent (Fig. 6.6b). Standard wind analyses have the same defect as conventional streamflow statistics—the time intervals are fixed in advance. Shellard's (1968) data, however, include hourly wind speeds which are useful in forecasting surface waves in small to medium sized lakes and which can be subject to the same extreme value analysis as floods. The mean annual hourly maximum, $\bar{W}_{hm}$, is the mean of a series of maximum hourly winds in each year and can also be estimated from the mean wind speed (Fig. 6.7a). The increase in wind speed with return period appears consistent throughout the UK.

6.2.3 Wind over Water

The surface roughness and the exposure of an anemometer on land is clearly different from one at the same height over water so that

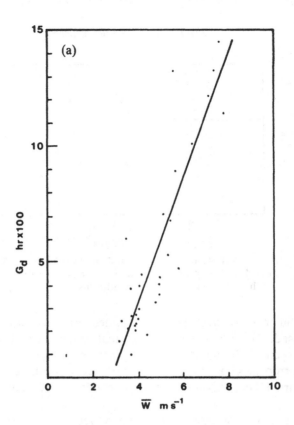

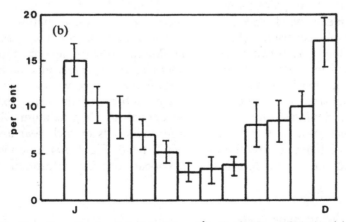

Fig. 6.6 Duration of gales (winds $> 17 \text{ m s}^{-1}$), G_d, in Great Britain. (a) Mean annual total duration of gales in relation to mean wind speed, $\bar{W}$, i.e. $G_\text{d} = 272 \cdot 5 \bar{W} - 765$. (b) Seasonal distribution, i.e., the percentage of that total in each month. Vertical bars indicate two standard deviations.

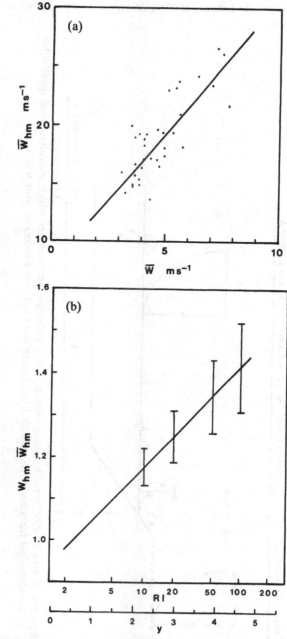

Fig. 6.7 Hourly maximum wind speeds, W_{hm}, in Great Britain. (a) Relation between the mean annual hourly maximum wind speed, $\bar{W}_{hm}$, and the mean wind speed, $\bar{W}$, i.e., $\bar{W}_{hm} = 2{\cdot}31\bar{W} + 7{\cdot}8$. (b) Relative increase in hourly maximum wind speed with recurrence interval. Vertical bars represent two standard deviations.

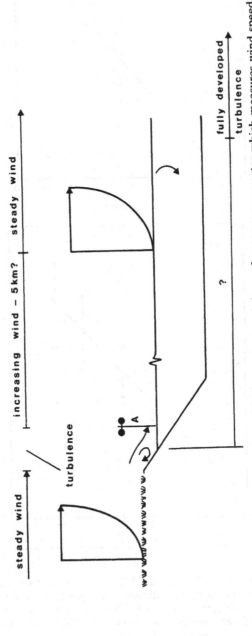

Fig. 6.8 Development of steady wind conditions over water. A is a reference anemometer which measures wind speed after the shoreline turbulence has subsided but before the wind speed increases over water.

differences in wind speed, for the same atmospheric conditions, are inevitable. The relation between the two wind speeds can be further complicated by turbulence near the lake shore, temperature differences between air and water and the distance required to establish steady wind conditions over water (see Fig. 6.8). The influence of topography on surface winds is a complex meteorological problem and no simple relation exists linking land and water wind conditions (see, for example, Richards *et al.* (1966) and Weringa (1986)). The data of Norman (1964) refer only to the increase of wind speed over water. He shows that, in comparison to winds measured near the shore, speeds increase for up to 10 km downwind, reaching 1·31 times the upwind value on average. Most of the increase, however, occurs in the first 5 km. Since Norman's reference anemometer is likely to be more exposed than a standard land based one, the ratio between land and water speeds is likely to be higher than the above.

Where the object is to estimate the mean wind speed in order to use the generalisations of Section 6.2.2, direct comparison of wind speed over water with a land based station is obviously desirable. A simple run-of-wind anemometer, read daily, is adequate. The accuracy of such correlations inevitably declines as the time interval decreases. Wind observations which are being related to current velocity, for example, must be made over water at the same time. It must be emphasised that the distance required to establish a steady wind over water is not the same as that required to establish steady conditions in the water itself. The conditions for fully developed turbulence in the water are rather uncertain and may, themselves, vary with wind speed (see Section 6.4.1).

6.3 THE HYDRAULIC STRUCTURE OF LAKES

The hydraulic structure of a water body can be envisaged as the spatial variation in its hydrodynamic characteristics. This is so simple in the case of rivers that the notion is hardly worth considering. Virtually the entire river is a gravity driven boundary layer flow whose characteristics are determined by the nature of the bed, the depth of water and the slope. Except in unusual circumstances, wind action is unimportant and the water is so well mixed that there are no temperature or density gradients within the flow and free turbulent interfaces only occur below the confluence of streams of different density.

Lakes, with their irregular form, greater depth and variety of forces causing motion, can have a much more complex hydraulic structure, i.e. within a lake there are water masses with different mean motions, some of it being almost steady and some definitely periodic. Vertical mixing is often weak so that stratification—the occurrence of layers of different temperature and density—is a common feature. Each of these water masses has its characteristic scales and intensity of turbulence. An essential feature of the hydraulic structure is that it is not fixed but is itself created by the same processes as determine the hydrodynamic characteristics of the separate water masses. There is an obvious analogy with meteorology. The weather we experience is the direct result of the air masses above us.

The forces causing motion in lakes are, obviously, the same as those that create the hydroclimate—sun, wind and water flowing off the land. Solar heating, when compared to the intensity of vertical mixing, determines whether a water body is stratified. Lateral differences in heating can cause horizontal density gradients that generate water movement but this effect is, normally, small compared to other forms of motion and is not considered further. Most commonly, wind is the primary cause of lake motion. Energy is transferred from the wind to the water, the water is set in motion and surface waves are generated. Inflowing rivers also add energy to a lake but, unless the lake is small or the outflow close to the inflow, river induced currents are usually lost in the wind driven motion.

One feature of the hydraulic structure has already been introduced in the discussion of turbulence (Section 3.2.1), i.e. the constant stress, Ekman and bottom boundary layers. More vertical detail can be added by considering possible stratification within the Ekman layer and the motion associated with surface waves. Horizontal structure can be due to upwelling and downwelling zones, irregular lake form, shore zones where there is interaction between surface waves and the lake bed, areas around inflows where the inflow current retains its identity and the region of convergent flow near the lake outlet. Some hydrodynamic analysis is necessary before the dimensions and characteristics of the hydraulic structure can be identified.

The hydrodynamics of lakes are complex and far from being fully understood. Heaps (1984) reviews the basic theory of motion in lakes. Extending the meteorological analogy, he provides the analytical basis for the equivalent of numerical weather forecasting. Imberger and Hamblin (1982) provide a review of the principal observational and theoretical work that has been done on lakes. Both are written for specialists and neither can be said to be easy to follow. What is attempted here is essentially descriptive—an account of the main features of motion in lakes. The emphasis is on elementary physics but with sufficient quantitative material included to assess the relative importance of the different phenomena that occur.

6.4 WIND–WATER INTERACTIONS

6.4.1 Basic Processes

Wind blowing over a water surface and the resulting wave formation is the most obvious example of free turbulent motion. Breaking waves transfer momentum to the water and it is easily seen from dimensional analysis that the transfer of momentum per unit area and time is equal to the stress (force per unit area) exerted by the wind on the water surface. As a result of this transfer process, both mean motion and turbulence is generated in the upper layers and the downward transport of kinetic energy creates drift currents at lower levels.

These are the basic processes that create wind driven motion in lakes. Further dimensional analysis shows that the energy transferred per unit area and time, i.e. the power per unit area derived from wind, P_w (W m^{-2}) is the product of the stress, τ_s, and velocity, U_s, at the surface, i.e.

$$P_w = \tau_s U_s \qquad (6.4.1)$$

Theoretical attempts to partition this power between wave formation, surface motion and the downward transport of kinetic energy have not proved fully successful so far (Fischer *et al.*, 1979). Empirical data must be used. This raises problems since any on-lake observations are the combined result of all processes acting, for example, an observation of current velocity includes the effect of wave action on the motion. It is well known that waves grow with fetch but it is usually assumed that, once wind edge effects have been taken into account, other wind related variables such as stress and current speed are not influenced by fetch. Imberger and Hamblin (1982), reviewing outstanding problems in lake hydrodynamics, indicate that these basic wind–water interactions, including the influence of fetch, need to be specified more accurately. Hansen (1978) emphasises the uncertainty about the length scale required for fully developed turbulence to occur in the water.

6.4.2 Empirical Observations

The fact that what happens at a boundary in fluid flow depends on the relative dimensions of the surface roughness and the laminar sublayer has already been discussed (Section 3.2.2). A number of observations indicate that the same distinction between smooth and rough flow occurs at the air–water interface. The most obvious example is the occurrence of white caps due to wave crests spilling over and foaming as described by Haines and Bryson (1961). These occur when the wind speed exceeds 4–5 m s^{-1}, irrespective of the size of the water body, i.e. they occur as a result of the properties of the interface and not as a consequence of a certain wave size occurring.

Deacon and Webb (1962) suggest the following relation between surface stress, τ_s, and W, the wind speed at 10 m above the water surface, viz. (Fig. 6.9)

$$\tau_s = (W^2 + 0.07W^3) \times 10^{-3} \qquad (6.4.2)$$

Also shown on Fig. 6.9 is the wind-stress relation assuming that the wind is blowing over a smooth surface. This supports the argument for a change in behaviour at wind speeds of 4–5 m s^{-1}. Equation (6.4.2) is only valid when the air and water temperatures are more or less the same but the stress increases when air is colder than water and vice versa. Bunker (1976) defines stress in terms of a drag coefficient, C_D, i.e. $\tau_s = \rho C_D W^2$, and tabulates C_D values for both varying wind speed and air–water temperature differences.

Establishing the relation between wind and surface current speed, U_s, is confounded by uncertainty in the definition of surface current. Smith (1979), referring to a thin layer near the surface, argues for a complex relation, identifying smooth, transitional and rough conditions (see Fig. 6.9). For fully rough conditions ($W > 10$ m s^{-1}), $U_s = 0.015W$. For lower wind speeds, the relation is equivalent to the polynomial

$$U_s = 0.014\,62 + 0.053\,49W - 0.008\,26W^2 + 0.000\,42W^3 \quad (6.4.3)$$

As shown below, there are associated changes with wind speed in the downward transfer of kinetic energy. In smooth conditions, surface velocities are relatively high but there is a rapid decline in velocity

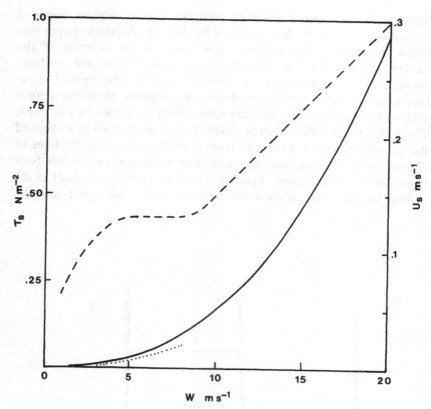

Fig. 6.9 Wind–water interactions. Full line is the relation between the wind speed, W, and surface stress, τ_s, (eqn (6.4.2)), the dotted line being the relation assuming that the water surface is smooth. Dashed line is the relation between wind speed and surface current velocity, U_s.

with depth while, in transition conditions, the constant surface velocity is balanced by increasing velocities at depth. The interaction between surface velocity and downward energy transport raises some uncertainty about the definition of surface current velocity to be used in estimating the power transferred by the wind. The surface layer velocity of Fig. 6.9 may not be appropriate.

6.4.3 Langmuir Circulations

An obvious feature of wind action on a lake is the appearance of streaks of foam, more or less orientated in the direction of the wind.

That these are associated with cellular vortex motion about a horizontal axis can be demonstrated by floating absorbent paper just below the surface. For a time, these travel in the direction of the surface current but, at intervals, they dive below the surface, subsequently re-appearing in a different position. The general structure of the motion, termed windrows or Langmuir circulations since they were first described by Langmuir (1938), is shown in Fig. 6.10. The visible foam, believed to be derived from oils formed as a result of the decomposition of plankton (Ewing, 1950), collects in the lines of convergence. The mechanisms generating such motion are not fully understood. Observations reported by Myer (1969) and Scott *et al.* (1969) indicate that streaks occur when the wind speed exceeds

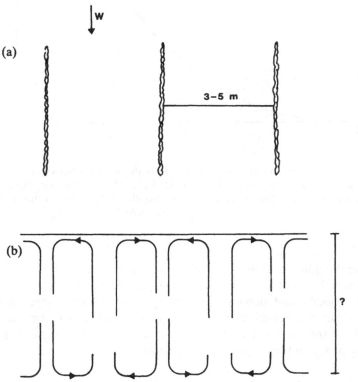

Fig. 6.10   General form of Langmuir circulations. (a) Plan view of water surface; (b) section through water surface.

$3 \, \text{m s}^{-1}$, that the streak spacing is generally between 3 and 5 m but may exceed 20 m and that the downwelling speed is a little less than 1 per cent of the wind speed for winds up to $10 \, \text{m s}^{-1}$. The rotations are asymmetrical, the upwelling velocity being about a third of that of the downwelling. There is some argument about the depth to which the cellular motion extends. Scott *et al.* (1969) indicate that the depth (m) is numerically the same as the wind speed (m s^{-1}) but others (see Buranathanitt *et al.*, 1982) suggest the motion extends down to the thermocline in stratified lakes and may reach the bed in shallow water. Some of the motion at the surface, therefore, has a definite structure and cannot be envisaged simply as mean motion with superimposed velocity fluctuations.

6.4.4 Hydrostatic Balance

There must be an equal and opposite force to that caused by the wind stress on the surface otherwise the lake would move bodily in the direction of the wind. For unstratified lakes, this balancing force is provided by the water surface tilting so that there is a hydrostatic pressure difference between the ends of the lake (Fig. 6.11). Currents near the bed are weak and the stress on the bed can usually be neglected. Equating the total force per unit width exerted by the wind, $\tau_s L_L$, to the pressure difference—the difference in the areas of triangles 1 and 2—an equation for the water surface slope, i, is obtained, i.e.,

$$i = \Delta D / L_L = \tau_s / \rho g D \qquad (6.4.4)$$

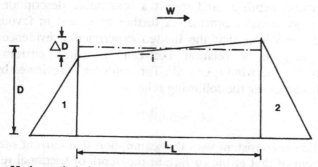

Fig. 6.11 Hydrostatic balance in an isothermal lake. The stress applied over the length of the lake is balanced by the hydrostatic force due to the difference in water level at the ends.

The force balance for stratified lakes is more complex and is discussed in Section 6.6.3.

6.4.5 Vertical Current Structure

The downward transport of kinetic energy generates currents at depths below the surface. To obtain the velocity depth profile, the argument favoured in some engineering literature emphasises mixing length theory and the fact that a wind driven current is equivalent to an inverted boundary layer flow. The differential equation for the velocity profile (3.2.9), therefore, applies, i.e.

$$dU = -U_f/k \,.\, dz/z \quad \text{or} \quad U = -U_f/k \,.\, \ln z + C' \qquad (6.4.5)$$

Given the boundary condition that $U = U_s$ at $z = 0$, the equation of Bye (1965) is obtained:

$$U = U_s - U_f/k \,.\, \ln z/z_o \qquad (6.4.6)$$

An alternative hypothesis assumes an exponential decline in wind drift velocity with depth, i.e.

$$U = U_s e^{-k_* z} \qquad (6.4.7)$$

There are several arguments in favour of an exponential profile. It can be seen as a simplification of Ekman's (1905) classic analysis of motion in the ocean. In terms of elementary physics, an exponential decline assumes that the rate of change of velocity, dU/dz, is proportional to the velocity at depth z, and this is a reasonable description of the kinetic energy transfer process. A further argument in favour of an exponential profile is that the limited experimental evidence that is available suggests a relation between the drift current decay coefficient, k_*, and wind speed, W. This evidence is reviewed by Smith (1979) who proposes the following relation

$$k_* = 6 \cdot 0 W^{-1 \cdot 84} \qquad (6.4.8)$$

This relation is consistent with the assumption that current speeds fall to 1 per cent of that at the surface at the depth of frictional resistance (see Section 6.5.1). There is an implication, therefore, that the coefficient may vary with latitude.

6.5 MOTION IN ISOTHERMAL LAKES

6.5.1 Introduction

Given the nature of wind, the notion of a steady state circulation has to be regarded as an idealised oversimplification. Much of the time, a lake is responding to rising wind or contains residual motion generated by previous wind. In reality, therefore, steady state circulation implies motion where the non-oscillating currents are stronger than those with a definite periodicity. Confirming what is already known about lake motion, we have

—the vertical division of a lake into the surface mixed layer, the Ekman layer and the bottom boundary layer;
—stress on the water surface balanced by a difference in water level between the ends;
—wind induced currents that decrease with depth.

What are the consequences of the above and what are the other constraints on the nature of the motion? The first essential is to meet the continuity requirement—in any volume within the lake, the quantity of water leaving must be balanced by the same quantity entering. The existence of a slope on the water surface generates gradient currents similar to those in a river. In most cases, the circulation pattern is determined by the way the continuity requirement and gradient currents interact, given the geometry of the lake. The third factor to be considered is that the currents may be weak enough for the effect of the earth's rotation to be significant.

All motion takes place on a rotating earth but any analysis relates to a fixed frame of reference. Because of the direction of the earth's rotation, compatibility between the two reference systems is achieved by assuming the existence of a small force acting to the right of the motion in the Northern Hemisphere and to the left in the Southern. This force is proportional to latitude and is zero at the equator. This Coriolis force, as it is often called, acts on all forces of motion, including that of express trains, but only when the other forces acting are of the same order of magnitude can its existence be detected.

Ekman's (1905) analysis of water movement in the ocean results in the classic Ekman spiral (Fig. 6.12). The Coriolis force causes the surface current to flow at 45° to the wind direction at the surface, the current speed declines exponentially with depth and the angle between

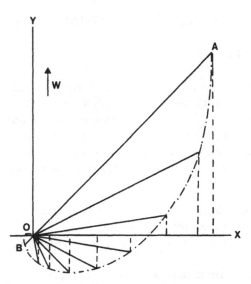

Fig. 6.12 Idealised motion in an Ekman spiral. Line OA is the surface current vector. Line OB is the current vector at the depth of frictional resistance. Curve ABO is the envelope of all the current vectors.

wind and current increases. At some depth, the inappropriately named depth of frictional resistance, D_*, the current flows in the opposite direction to that of the wind and its speed is small. D_* can be calculated from the following equation

$$D_* = \pi \tau_s / \sqrt{2} \, \rho \Omega \quad U_s \sin \phi \qquad (6.5.1)$$

where Ω is the angular velocity of the earth's rotation ($7 \cdot 29 \times 10^{-5}$ rad s^{-1}) and ϕ is the latitude.

Whether or not the earth's rotation affects the motion in a lake depends on its depth and lateral dimensions as well as the wind speed. In a shallow lake, the energy transferred to the water from the wind is compressed into a lesser depth and the forces acting are large compared to the Coriolis force. This effect can be measured by D/D_*. The influence of lateral dimensions is measured by the Rossby Number, R_o (Fischer *et al.*, 1979) where

$$R_o = U/L'\Omega \qquad (6.5.2)$$

where U and L' are a characteristic velocity and length dimension. The characteristic length dimension is, perhaps, best replaced by $\sqrt{A_L}$

where A_L is the surface area of the lake. The larger the R_o value, the more likely it is that the fluid momentum will dominate over the effect of the earth's rotation. Neither D/D_* nor R_o can be considered as giving precise estimates of the influence of the Coriolis force and no critical values can be quoted with any confidence.

6.5.2 Two Dimensional Steady Circulation

The simplest form of circulation that meets the continuity requirement is a two dimensional circulation where the effect of the earth's rotation is neglected. The water transport in the direction of the wind due to the drift current is exactly balanced by a gradient flow in the opposite direction. An analytical solution of this problem is given by Banks (1975) and indicates the velocity at any depth, z, in terms of a characteristic velocity, U_*, i.e.

$$U/U_* = (1 - z/D) - 3/4[1 - (z/D)^2] \tag{6.5.3}$$

where D is the total water depth (see Fig. 6.13). The net forward and back water transports are, obviously, the same and equal to $4U_s D/27$. By setting $z = 0$, $U_s/U_* = 1/4$. A feature of eqn (6.5.3) is that the profile form is the same at all wind speeds, only the scale being changed. The extent of the upwelling and downwelling zones is not known.

The solution to the same problem by Smith (1979) is, analytically, much less tidy. The exponential drift current and the gradient current are treated separately so that the velocity at any depth is the difference between the two, i.e.

$$U = U_s e^{-k_* z} - 5 \cdot 75 U_f \log[30(D - z)/r_p] \tag{6.5.4}$$

The second term on the right hand side is the standard equation for gradient driven flow (Section 3.2.2). A tedious numerical procedure is necessary to determine the depth of zero velocity and U_f, the friction velocity for the gradient current, that ensure that the net forward and back transports are equal. The advantage of eqn (6.5.4) is that the known dependence on wind speed of the coefficient, k_*, can be incorporated so that the profile form changes appreciably with wind speed (see Fig. 6.14). This is believed to be closer to what actually occurs.

If the effect of the earth's rotation can be detected and currents are deflected, velocities can be resolved into components in the direction

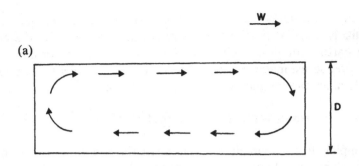

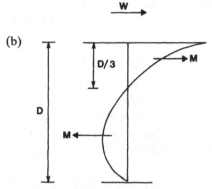

Fig. 6.13 Two dimensional circulation in an isothermal lake (Banks, 1975). (a) General form of circulation; (b) velocity profile at the centre of the lake, the forward and return water transports, *M*, being equal.

of and at right angles to the wind direction. The circulation in the direction of the wind is, essentially, as described above. The existence of velocity components at right angles to the wind causes a slight rise in water level on the bank and the continuity requirement must also be met in the transverse direction, resulting in a weak cross circulation. The idealised net result is that the trajectory of a water particle is helical. Evidence of this form of motion in the epilimnion of a stratified lake was obtained by George (1981) although observations were not made at sufficient depths to confirm the transport balance (see Section 6.6.5).

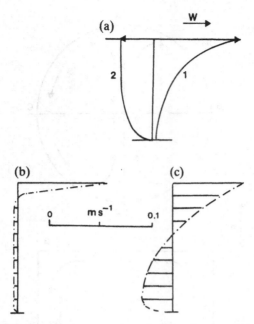

Fig. 6.14 Velocity profiles in a two dimensional, isothermal circulation (Smith, 1979). (a) Opposing wind drift (1) and gradient (2) currents, i.e., eqn (6.5.4); (b) combined current in lake 20 m deep subject to a wind of $2\,\mathrm{m\,s^{-1}}$; (c) the same subject to a wind of $10\,\mathrm{m\,s^{-1}}$.

6.5.3 Circulation in Irregularly Shaped Lakes

The form of circulation pattern occurring in irregularly shaped lakes is difficult to predict from simple physical arguments. Instead of the simple tilt of the water surface as in the previous two dimensional circulation, the pattern of water level distortion can be quite elaborate and how the resulting gradient currents combine with the continuity requirement is rarely clear. One idealised pattern in a lake of constant depth is that described by Livingstone (1954) which can be reproduced by blowing over a soup plate (see Fig. 6.15a). The water level will be highest at A so that longshore gradient currents towards B and C can occur. The does not preclude the existence of a bottom return current but the conditions for transport balance are now altered. Since water surface slope is inversely proportional to depth (eqn (6.4.4)), depth differences alone can affect the circulation pattern. In Fig. 6.15b, water level will be highest at A and some flow towards B can occur so

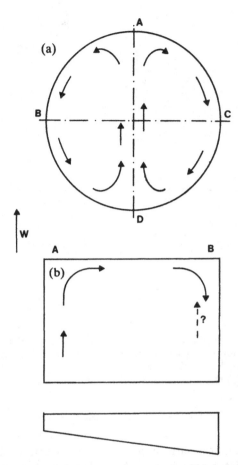

Fig. 6.15 Schematic circulation patterns in irregularly shaped lakes. (a) Circular lake of uniform depth; (b) the effect of depth variations (see text for details).

that again the conditions for transport balance are unclear. Equation (6.5.4) can also be used to give the velocity profile for different ratios of forward and return transport (see Fig. 6.16).

Figure 6.17 shows the observed surface current directions as well as the near bottom current directions in two deeps in Loch Leven (Smith, 1974). The flow of surface water towards the South East corner is much greater than the discharge down the outflow, there is a downwelling where no detectable velocity is indicated and the con-

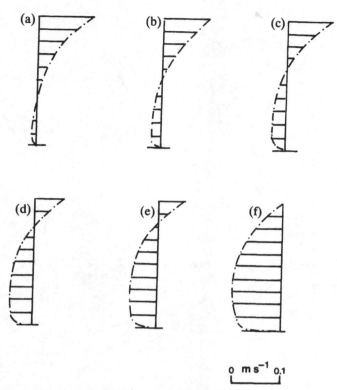

Fig. 6.16 Velocity–depth profiles for different ratios of forward to return transport in a lake 10 m deep subject to a wind speed of $5\,\mathrm{m\,s^{-1}}$. (a) 10; (b) 2; (c) 1; (d) 0·25; (e) 0·10; (f) zero velocity at surface.

tinuity requirement is met by a return flow through the trough to the South of St Serf's Island. Horizontal rotations with the surface current running against the wind, akin to the Livingstone type circulation, are seen to the North West of St Serf's Island.

6.5.4 Surface Seiches

The horizontal force applied to a lake by the wind stress is, under 'steady' conditions, balanced by the hydrostatic pressure difference due to the tilt of the water surface. If the wind suddenly drops, that balance is destroyed, water rushes to the other end and the lake water starts oscillating. Darbyshire and Darbyshire (1957) show that oscillation can be induced without a complete cessation of wind—a combina-

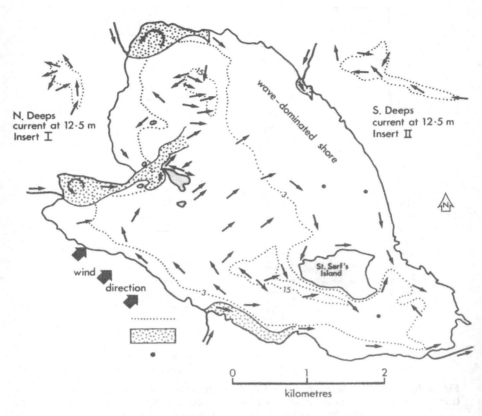

Fig. 6.17 Observed surface circulation pattern in Loch Leven (based on Smith, 1974). Each arrow represents a drogue observation. The dotted lines are underwater contours, the stippled areas show zones coloured by river water at times of high inflow and the dots indicate where there was no detectable velocity. The insets show observed current directions at depth.

tion of change in speed and direction is enough. Such oscillations or surface seiches can have other causes such as the sudden operation of a hydro-electric plant but disruption of the hydrostatic balance is the most usual.

Observation of water level is easy compared to current velocity and the theory of seiches is based on classical mathematical physics rather than on more modern fluid dynamics so that seiches were one of the first topics investigated in lake hydrodynamics. Surface seiches, besides their intrinsic interest, shed some light on the meaning of steady state motion. Oscillation, once started, takes some time to decay so that Shulman and Bryson (1961), for example, made all their current observations over a time interval equal to the period of the primary oscillation so as to remove the effect of oscillating motion. This is not always possible and the existence of residual periodic motion, the need for a definite start up time to establish a circulation and the fact that, once established, inertial effects may mean that currents generated by antecedent wind conditions may still be running all contribute to the uncertainty in any observation of current velocity.

The essential features of seiche motion can be demonstrated by considering what happens in a rectangular basin. The oscillation has a wavelength, λ_s, which is, effectively, equal to twice the lake length, L_L (see Fig. 6.18). The position of zero displacement is termed the node and the motion is characterised by its amplitude, a_f, and period, T', the time for a complete cycle. The characteristic velocity of the oscillation, the wave celerity, c, is λ_s/T'. Since the wavelength is large compared to the water depth, the celerity is also equal to $\sqrt{gD}$ (Section 6.7.1) so that

$$T' = 2L_L/\sqrt{gD} \qquad (6.5.5)$$

At the instant of zero displacement, all the energy of the oscillation is kinetic. By equating the total potential energy due to the surface tilt to the maximum kinetic energy, an equation for the maximum current speed of the gradient flow, u_m, can be obtained, i.e.

$$u_m = \Delta D\sqrt{g/D} \qquad (6.5.6)$$

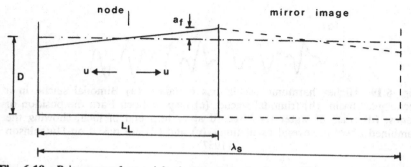

Fig. 6.18 Primary surface seiche in a rectangular basin. The oscillation of the surface can be seen as wave motion where the wavelength is equal to twice the lake length.

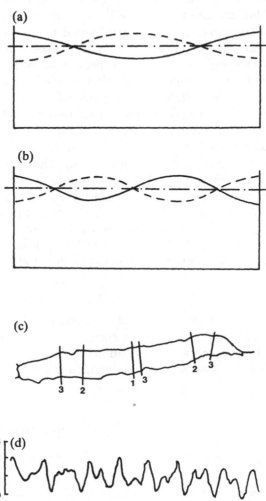

Fig. 6.19 Higher harmonic oscillations in lakes. (a) Bimodal seiche in a rectangular basin; (b) trimodal seiche; (c) map of Loch Earn the position of nodes; (d) trace of water level record at eastern end of loch, showing the combined effect of several oscillations. (c) and (d) are based on Hutchinson (1957).

A feature of seiches, even in basins of regular shape, is the tendency for secondary oscillations, having a number of nodes, to be formed (Fig. 6.19). The occurrence and intensity of these higher harmonic oscillations cannot be predicted in advance. Their existence cannot create energy, however, so that the original potential energy has to be distributed between the different harmonics. The period of a higher harmonic, having n nodes, is given by

$$T'_n = 1/n \, . \, 2L_L/\sqrt{gD} \tag{6.5.7}$$

As with most simple oscillations, the amplitude of a seiche decays exponentially, the period being virtually unaffected. In the case of the primary seiche, Shiau and Rumer (1974) suggest the following equation for the maximum amplitude at time t, a_{ft},

$$a_{ft} = a_{fo} e^{-k_s t/T'} \tag{6.5.8}$$

where a_{fo} is the initial maximum amplitude, T' is the period and k_s is the decay coefficient. The energy loss, and hence the magnitude of the decay coefficient, is caused by the generation of turbulence in the water and, particularly in shallow lakes, by frictional losses at the bed.

The theory of surface seiches in small to medium sized lakes is fully discussed by Hutchinson (1957) who also considers the complex oscillations that can occur in basins of irregular shape and varying depth. A numerical procedure for calculating seiche characteristics in complex water bodies is given by Defant (1918, 1961) but see also Lemmin and Mortimer (1986). In large water bodies where the effect of the earth's rotation is pronounced, rotating oscillations, first described by Mortimer (1963) can occur. A full review of all forms of oscillation in lakes is provided by Hutter (1984).

6.6 MOTION IN STRATIFIED LAKES

6.6.1 The Occurrence of Stratification

If a still water body is subjected to radiant heat, the surface cooling due to evaporation sets up convection currents which mix the upper layers. Instead of a continuous vertical temperature gradient occurring, the water is divided into a circulating, heated upper layer on top of a cold, stagnant layer. Gentle stirring by wind tends to increase this separation. The result is the familiar vertical temperature structure of a lake, i.e. the warm, turbulent epilimnion above the cold, relatively undisturbed hypolimnion, the two separated by a zone of strong temperature gradient, the thermocline (see Fig. 6.20).

Vigorous mixing, however, destroys this vertical structure and the whole lake is heated to the same temperature. The criterion that determines whether the turbulence is sufficiently intense to destroy the stratification or not is the Richardson number, R_i. This is defined as the ratio of the work that has to be done to lift a volume of water against the density gradient to the available turbulent kinetic energy. For a density gradient, $d\rho/dz$, and a velocity gradient, dU/dz, R_i is given by (Smith, 1975)

$$R_i = (g \, d\rho/dz)/(dU/dz)^2 \qquad (6.6.1)$$

Whenever the work to be done is greater than the available energy, the turbulence will die away rapidly. Even if the work to be done is less than the total available energy, some energy will be consumed in raising water against the gradient so that the turbulence will be progressively weakened. It is to be expected, therefore, that the critical value of R_i for the ultimate disappearance of turbulence will be less than one. The actual value is about 0·25 (see, for example, Miles, 1963).

The data presented by Straskraba (1980) suggest the following rough rule for the depth of the thermocline, z_e, in terms of the lake length, L_L (km), viz.

$$z_e = 4\sqrt{L_L} \qquad (6.6.2)$$

There is some indication that, in exposed areas, the coefficient is greater than 4 and vice versa in sheltered sites. The equation gives a guide as to whether stratification will occur. If the predicted thermocline depth is greater than the maximum depth of the lake, then, obviously, stratified conditions are unlikely while, if the predicted depth is less than the maximum depth but greater than the mean depth, intermittent stratification in deep areas may occur.

Equation (6.6.2) can only be a very rough guide since the depth of the thermocline increases with time. Analysis of temperature data (Ramsbottom, personal communication) for the South Basin of Windermere indicates that

$$z_e = 0·07t + 10·2 \qquad (r = 0·781) \qquad (6.6.3)$$

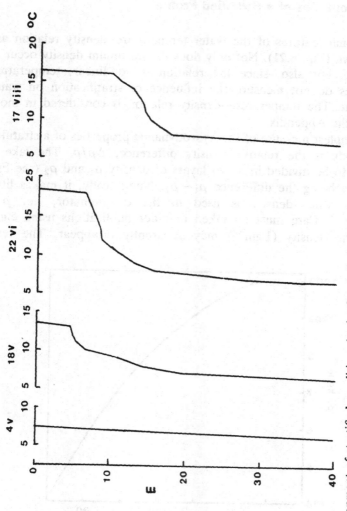

Fig. 6.20 Development of stratified conditions in the South Basin of Windermere. The numbers at the top are the dates of observation.

where t is the number of days since the onset of stratification. The deepening does not occur at a steady rate but intermittently during periods of high wind. Density stratification can be caused by other than temperature but this is not considered.

6.6.2 Properties of a Stratified Profile

The unusual features of the water temperature–density relation are well known (Fig. 6.21). Not only does the maximum density occur at about 4°C but also, since the relation is non-linear, temperature differences do not measure the influence of stratification on water movement. The temperature–density relation is considered in more detail in the Appendix.

The simplest measure of the hydrodynamic properties of a stratified water body is the relative density difference, $\Delta\rho/\rho$. The lake is assumed to be divided into two layers of density ρ_1 and ρ_2 (see Fig. 6.22), $\Delta\rho$ being the difference $\rho_1 - \rho_2$. Numerically, it makes little difference what density is used in the denominator, i.e. $\rho = 1000$ kg m^{-3}. Care must be taken in older publications using c.g.s. units. The density (1 gm^{-3}) may apparently disappear. The term

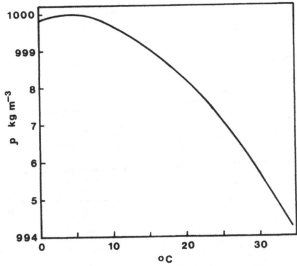

Fig. 6.21 Relation between density, ρ, and temperature for pure water (from Hutchinson, 1957).

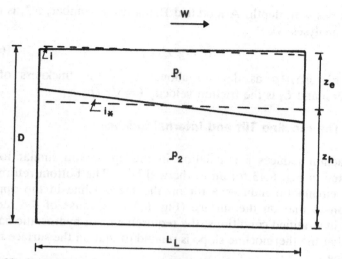

Fig. 6.22 Definition sketch for stratified motion assuming a two-layered structure.

$g\Delta\rho/\rho = g'$ expresses the reduced gravitational acceleration in stratified flow. Figure 6.20 shows that the assumption of a two layered structure can be a considerable oversimplification but analysis of three layered systems presents considerable difficulty. Mortimer (1953) suggests a method for identifying the two layered structure given the temperature profile and lake morphometry.

An alternative measure of profile characteristics is the Brunt–Vaisala frequency. If, in the presence of a density gradient, a particle of water is displaced from its equilibrium position and then released, its motion is similar to that of a weighted spring. The frequency of this oscillation is the Brunt–Vaisala frequency, N_*, where

$$N_* = \sqrt{g/\rho \cdot d\rho/dz} \qquad (6.6.4)$$

It is easily seen that

$$R_i = N_*^2/(dU/dz)^2$$

Both the relative density differences and the Brunt–Vaisala frequency are static measures that do not incorporate any indication of water movement. The Richardson Number, as defined by eqn (6.6.1), has two disadvantages—the velocity profile must be known and its value varies with depth. A modified Richardson Number, R_i^*, is often used in analyses, viz.

$$R_i^* = g'z_e/U_f^2 \tag{6.6.5}$$

where $g' = g\Delta\rho/\rho$ as defined above, z_e is the thickness of the epilimnion and U_f is the friction velocity ($= \sqrt{\tau_s/\rho}$).

6.6.3 Thermocline Tilt and Internal Seiches

Wind action induces a circulation in the epilimnion similar to that illustrated in Fig. 6.13 for an isothermal lake. The bottom return flow of this circulation induces a tilt on the thermocline in the opposite direction to that on the surface (Fig. 6.22). Because of the reduced gravity in stratified conditions, the requirements of hydrostatic balance mean that the thermocline slope is related to that on the surface by the following

$$i_* = i\rho/\Delta\rho \tag{6.6.6}$$

Since $i = \tau_s/\rho gD$ (eqn (6.4.4)) and replacing D by z_e,

$$i_* = 1/R_i^* \tag{6.6.7}$$

As with the surface seiche, any change in the hydrostatic balance will cause the hypolimnion to oscillate, resulting in an internal seiche. The period of the primary internal seiche, T'_*, is given by

$$T'_* = 2L_L(g'Z_eZ_h/D)^{-1/2} \tag{6.6.8}$$

An estimate of the maximum current velocity associated with internal oscillations is given by Spiegel and Imberger (1980), i.e.

$$u'_m = U_f^2 T'_*/4z_e \tag{6.6.9}$$

The traditional view of internal seiches emphasises the simple to and fro motion associated with a dominant primary oscillation, i.e. higher harmonics are weak, if present at all, and the only motion in the hypolimnion is the oscillating gradient current. More recent research has shown, as is almost inevitable, that this view is too simple. Thorpe (1971, 1972) demonstrates that an internal seiche may not involve simple oscillation of the surface, the observed motion in Loch Ness being asymmetrical and more like that of an internal bore or surge. LaZerte (1980) emphasises that the assumption of a two layered structure can lead to serious error. Where the thermocline is thick compared to the other layers, complex oscillations involving higher harmonics predominate. Imberger and Hamblin (1982), reviewing the possible application of oceanographic research to lakes, suggest that complex wave motions can interact with each other and the lake boundaries, resulting in the sporadic generation of patches of turbulent motion in the hypolimnion. Arguments for the existence of bursts of more intense motion are supported by observations of current velocity. Lathbury *et al.* (1960) observed currents in the hypolimnion of Lake Mendota that could not be explained by the gradient current of the seiche motion. Lemmin and Imboden (1987) observed bottom currents sufficiently intense to cause particle resuspension. Interactions between previous and newly generated internal seiche motions are observed by Horn *et al.* (1986).

Lemmin and Mortimer (1986) extend Defant's method for calculating surface seiche characteristics for real morphology to internal seiches. The method allows displacements, and hence current speeds, to be estimated and a modification, taking some account of the influence of the earth's rotation, is included. The implication of the above, however, is that sporadic bursts of more intense motion can occur. These may not be predictable but could have important consequences for exchange processes at the sediment–water interface.

6.6.4 Regimes of Stratified Motion

Progressive deepening of the thermocline has already been demonstrated but the physical processes causing this deepening are the subject of controversy (Spiegel and Imberger, 1980). The possible mechanisms proposed are stirring caused by the downward transport of kinetic energy from the surface and shear flow, velocity gradients, in the vicinity of the interface which erodes the top of the lower layer. The rate of thermocline deepening, dz_e/dt, can be considered as an entrainment velocity, E_v. Blanton (1973) shows that E_v, measured as the change in epilimnion volume over a time interval divided by the lake surface area, increases with mean depth. He also gives an empirical relation between E_v and a stability measure, $S_* = g/\rho \cdot \Delta\rho/\Delta z$, where $\Delta\rho/\Delta z$ is the density gradient over the thermocline. The relation is the inverse of that required to predict the entrainment rate, i.e.

$$S_* = 8 \cdot 95 \times 10^{-7}/E_v^{0.521} \qquad (6.6.10)$$

A further feature of stratified motion is that it is an example of free turbulence (see Section 3.2.5). Some form of wave formation and Kelvin–Helmholtz instability at the interface may be expected to occur. The classic account of such wave formation in lakes is that given by Mortimer (1961). Thorpe (1971) directly observed what he termed billows in Loch Ness—wave like structures about 2·7 m long and 0·8 m high. It is reasonable to expect that wave formation would be more pronounced when velocity gradients are greater—the height of wind generated waves on the surface increases with wind speed.

Three features of stratified motion, therefore, have been identified: thermocline tilt and internal oscillations; progressive thermocline deepening; wave formation on the interface. How do these processes interact and do they all occur at the same time? It is also necessary to know whether a particular state of motion persists under any given conditions. Only under persistent, stable conditions can anything approaching steady state circulation be considered.

Spiegel and Imberger (1980) present a classification of lake behaviour in which the relative importance of the different processes can be seen. The analysis is restricted to two layered, rectangular basins in which the effect of the earth's rotation can be neglected. The physical argument is based on identifying the processes that cause thermocline deepening and comparison of the various time scales involved, particularly the time required for the upper layer to deepen to the bottom, the entrainment time, and the period of the primary internal seiche. The characteristics of the motion in the different regimes are summarised in Table 6.2.

The evidence presented on the criteria that separate the various regimes is not totally consistent. The original scheme is based on the relation between the modified Richardson Number, R_i^*, and the aspect ratio of the lake, i.e. L_L/z_e. Later work (Patterson *et al.*, 1984) combines these groups of variables into the Wedderburn Number, W_*, where

$$W_* = g'z_e^2/U_f^2 L_L \qquad (6.6.11)$$

The change from one regime to another is not abrupt and the use of a single criterion measuring the characteristics of stratification is preferable. In general, the larger the value of W_*, the more stable the stratification and, when $W_* = 1$, the interface surfaces at the upwind end. The criteria for separating the regimes in Table 6.2 are from Patterson *et al.* (1984) and are derived from numerical experiments

Table 6.2 Regimes of stratified motion

Regime	Criterion	Characteristics of motion
1	$W_* \ll 1$	—basin is, effectively, homogeneous —vertical mixing is extremely rapid
2	$3 < W_* < 10$	—deepening occurs mainly due to velocity shear at interface —large interface displacements —presence of billows and a thick, unstable thermocline —internal seiches are not observed —complete mixing may occur during a single wind episode
3	$W_* > 10$	—deepening results from stirring from the surface —internal seiches are the dominant feature —interface is sharp and billows less pronounced
4	$W_* > 25?$	—weak internal oscillations —interface is sharp and billows are not observed —complete mixing does not occur

with a computer model. Actual lake data are given by Spiegel and Imberger (1980) and, from their Table 2, estimates of W_* obtained. The results are consistent in that the thermocline surfaces for the lowest values of W_* and the lakes in regime 3 have considerably higher values of W_* than those in regime 2. The numerical limits, however, do not coincide exactly with those in Table 6.2 but there must, inevitably, be considerable uncertainty in the estimated values. Regime 4 is only observed in a short laboratory tank with strong stratification and where the estimated value of W_* is about 34.

Additional difficulties stem from the recurrent problem with stratified motion, namely, that, where the thermocline is thick compared to the other layers, the assumption of a two layered structure is inadequate. Motion in a three layered lake is analysed by Monismith (1985) who demonstrates the additional complexities that can arise. Despite these uncertainties, continuous updating of the Wedderburn Number from wind and temperature profile data provides a useful indication of the state of stratified lakes.

The destruction of stratification and the return to mixed, isothermal conditions in temperate lakes is often referred to as the autumn overturn. The terminology is unfortunate in that it implies that the return to homogeneity is caused solely by surface cooling and a literal overturn because water at the surface is denser than that at a greater depth. Uniform conditions are more likely to result from the upper layer penetrating to the bottom, particularly as the value of the Wedderburn Number falls.

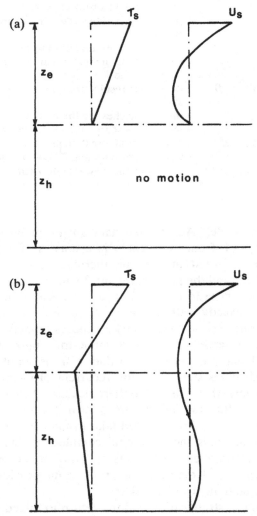

Fig. 6.23 Stress and velocity profiles for a two-dimensional circulation in a stratified lake. (a) Free slip, zero stress condition at the interface; (b) induced motion in the hypolimnion.

6.6.5 Steady State Circulation in Stratified Lakes

Given wind regimes that allow stratification to persist, the form of circulation occurring in stratified lakes depends on the properties of the interface, i.e. whether or not motion in the upper layer induces motion in the hypolimnion. Detailed analytical considerations are given by Heaps (1984). It is often assumed that there is free slip and zero stress at the interface so that there is no motion in the lower layer (Fig. 6.23a). If there is stress at the interface, then a reversed circulation in the hypolimnion occurs (Fig. 6.23b). The free slip condition implies that, in terms of the general circulation pattern, the interface is, effectively, a solid boundary and the possible circulation patterns in the epilimnion are virtually identical to those in an isothermal lake. Any induced circulation in the hypolimnion is likely to be weak and difficult to detect, given the induced motion from earlier, residual internal oscillations. Observations by George (1981) in the South Basin of Windermere, support the argument for treating the thermocline as a virtual lake bed as far as the general circulation is concerned. Observations were confined to the central region of the lake, however, and nothing is known about possible edge effects where the thermocline impinges on the lake bed. Blanton (1973) refers to 'leakage round the edge' as a possible mechanism for entraining hypolimnetic water into the epilimnion. Not only may this generate additional turbulence but also the nature of the upwelling and the transport balance of the circulation could be affected.

6.7 SURFACE WAVES

6.7.1 Wave Characteristics

Waves travelling on the lake surface are the most obvious feature of wave action. Such waves are characterised by their height, H, the difference between crest and trough, wavelength, λ, the distance between crests, and period T, the time interval between crests passing a fixed reference point (see Fig. 6.24). Since a wave travels a distance λ, in time T, the speed at which a wave crosses the surface, c, usually referred to as the wave celerity, is equal to λ/T. The formal, mathematical theory of wave motion (Lamb, 1945) gives the following equation for the wave celerity in water of depth D, i.e.

$$c = \sqrt{g\lambda/2\pi} \cdot \tanh 2\pi D/\lambda \qquad (6.7.1)$$

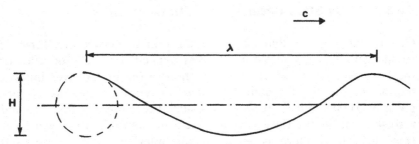

Fig. 6.24 The features of waves on a lake surface.

Associated with the passage of the wave is orbital motion of individual water particles. At the surface, the motion is virtually circular and is usually assumed to be closed, i.e. the particle returns to its original position in one complete period, the radius being equal to half the wave height. In reality, the particle orbit is not completely closed and some forward transport of water can be directly ascribed to wave action.

The nature of waves and of the orbital motion beneath them is determined by the value of $\tanh 2\pi D/\lambda$. When $D/\lambda = 0.5$, $\tanh 2\pi D/\lambda = 1$, so that

$$c = \sqrt{g\lambda/2\pi} \tag{6.7.2}$$

This critical value of D/λ indicates the limiting condition for the occurrence of deep water waves whose celerity and other characteristics are independent of depth. Since $\lambda = cT$,

$$\lambda = 1.56T^2 \tag{6.7.3}$$

When $D/\lambda = 0.05$, $\tanh 2\pi D/\lambda \doteqdot 2\pi D/\lambda$, so that

$$c = \sqrt{gD} \tag{6.7.4}$$

This is the condition for the occurrence of shallow water waves whose celerity is independent of wavelength. Within these two limits, transitional waves occur whose characteristics depend on both D and λ. Both shallow and transitional types can be referred to as shore zone waves, i.e. the motion associated with the waves extends to the lake bed and may influence bottom dwelling species. The book by King (1972), although concerned with ocean waves, gives a useful account of wave features without excessive mathematics.

Plate 13 Storm conditions and spindrift on Loch Ness.

Plate 14 Breaking waves on the shore of Loch Ness.

6.7.2 Deep Water Waves

Watching an individual wave shows that a crest appears to move through a group and then disappear—individual waves travel faster than the group of which they form a part. The wave celerity, as defined above, measures the rate of advance of an individual crest. Formal wave theory shows that this group velocity is equal to half the wave celerity.

The other obvious feature is that not all waves are the same size either in terms of height or length. The standard measure of height is usually taken to be the significant wave height, H_s, defined as the average height of the highest third of all the waves. Observed wave height distributions, standardised in terms of H_s, appear to be more or less constant (see Fig. 6.25). Similar arguments apply to the significant wave period, T_s. Subsequent discussion relates to H_s and T_s and wave variability is not considered.

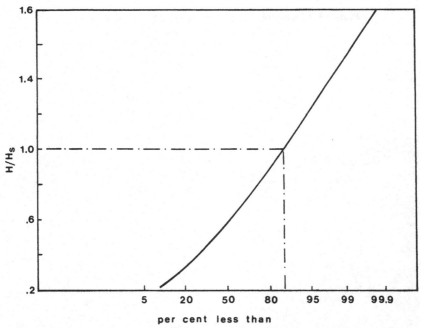

Fig. 6.25 Standardised wave height distribution curve. Wave height, *H*, is expressed as a proportion of the significant wave height, H_s. 87 per cent of waves are lower than the significant wave height.

Provided the wind is blowing for long enough, wave characteristics are determined by the wind speed, W, and the effective fetch, F. Prediction equations for this condition are (CERC, 1977)

$$gH_s/W^2 = 0.283 \tanh[0.012\,5(gF/W)^{0.42}] \qquad (6.7.5)$$

$$gT_s/2\pi W = 1.2 \tanh[0.077(gF/W)^{0.25}] \qquad (6.7.6)$$

That the fetch is not the greatest straight line distance upwind is obvious from the fact that large waves are not generated in long, straight canals. Exactly how to calculate the effective fetch is the subject of some debate. The original procedure (see, for example, Smith and Sinclair, 1972) is to draw seven radial lines at 6° intervals to either side of the axial wind direction, the length of each radial line multiplied by the cosine of the angle between the radial and the wind direction being summed and then divided by the sum of all the cosines (13·5). A later suggestion is that the radial lengths should be multiplied by the squares of the cosines but the original procedure is used here. A computer program for calculating effective fetch, using digitised map data, is given by Hilton and Rigg (1985). The influence of lake shape, expressed as shoreline development, on the ratio of the maximum effective fetch to the axial length of a lake is shown in Fig. 6.26. Wave development can also be inhibited by restricted water depth upwind.

Equations (6.7.5) and (6.7.6) are only valid if the wind blows for long enough for wave growth with time to cease. The minimum wind duration for steady state waves to occur, t_d, is (US Army, 1962)

$$t_d = 29.8F/T_s \qquad (6.7.7)$$

where t_d is in min, F in km and T_s is the wave period calculated from eqn (6.7.6). For typical wave periods of 1–2 s, t_d is equal to 60 min for fetches of around 4–8 km. This is the justification for assuming that hourly maximum wind speeds are appropriate for estimating severe wave conditions in small to medium sized lakes (see Section 6.2.2). In tropical lakes particularly, the actual storm duration may be less than t_d. Viner and Smith (1973) suggest the following equations for H_t and T_t, the significant wave length and period when the actual storm duration, t_a, is less than t_d

$$H_t/H_s = (8.95t_a/t_d)/(1 + 7.95t_a/t_d) \qquad (6.7.8)$$

$$T_t/T_s = (4.35t_a/t_d)/(1 + 3.35t_a/t_d) \qquad (6.7.9)$$

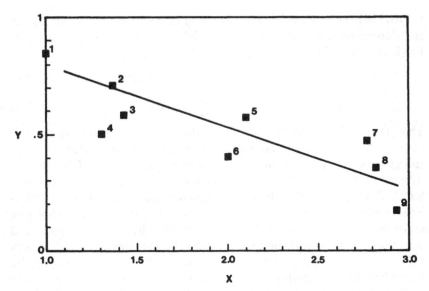

Fig. 6.26   The influence of shoreline development, X, on the ratio of effective fetch to axial length, Y. The best fit relation is $Y = 1{\cdot}065 - 0{\cdot}269X$. 1 circle; 2 Loch Insh; 3 Loch Leven; 4 Crosemere; 5 Lough Neagh; 6 Esthwaite Water; 7 Loch Lomond; 8 Windermere; 9 Loch Ness.

In deep water, the motion of water particles is circular at all depths, the radius, r_w, at depth z being given by

$$r_w = H_s/2e^{-2\pi z/\lambda} \qquad (6.7.10)$$

The time to complete one orbit is the period, T, so that the orbital velocity, v_w, is given by

$$v_w = H_s/T_s e^{-2\pi z/\lambda} \qquad (6.7.11)$$

This orbital motion can be envisaged as a form of 'turbulence' superimposed on any other motion, i.e. there is, effectively, a vertical velocity and an associated length. The depth of the wave mixed layer, z_m, is often taken to be equal to $\lambda/2$, largely for convenience since it is the same as the definition of deep water. Strictly, all that $z_m = \lambda/2$ implies is that the orbital radius is reduced to $e^{-\pi}$, i.e. 4 per cent of the surface wave height. The radius and velocity are not constant at this depth (see Smith and Sinclair, 1972).

6.7.3 Shore Zone Waves

As waves approach the shore and the water depth becomes less than half the wavelength, wave characteristics are altered. There is no immediate energy loss so that the energy associated with the wave motion is compressed into ever decreasing depth. The result is increased wave height and reduced wave length. Defining H_{so} and λ_{so} as the significant height and length at the shore zone limit, the characteristics inshore, H_o and λ_o, in water of depth D are given by the following

$$H_o/H_{so} = [\sinh(4\pi D/\lambda_{so})/[4\pi D/\lambda_{so}$$
$$+ \sinh(4\pi D/\lambda_{so}] \cdot 1/(\tanh 2\pi D/\lambda_{so}]^{1/2} \quad (6.7.12)$$

$$\lambda_o/\lambda_{so} = \tanh(2\pi D/\lambda_{so}) \quad (6.7.13)$$

The form of the transformations are shown in Fig. 6.27. Useful tables of these tedious functions are given by Raudkivi (1976). The wave

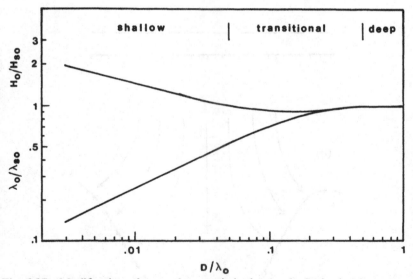

Fig. 6.27 Modification of wave characteristics by depth, D, in the shore zone. λ_{so} is the wavelength at the shore zone limit. When λ_{so}/D is greater than 0·5, deep water waves are unaffected by water depth. When λ_{so}/D is less than 0·5, the relative wave height, H_o/H_{so}, increases and the relative wave length, λ/λ_{so}, decreases. Shallow water waves occur when D/λ_{so} is less than 0·05. The wave period is unaffected by depth.

steepness, H_o/λ_o, increases until the waves become unstable and the crests topple over. As a rough rule, waves break when $H_o/D = 0.78$. Inshore of the breaker line, swash runs up and down the beach.

If waves approach the shore obliquely, the direction of wave travel is affected as well as height and length—a process usually referred to as wave refraction. The wave celerity is reduced at the same rate as the wavelength, i.e. $c/c_o = \lambda/\lambda_o$. At point A on Fig. 6.28a where

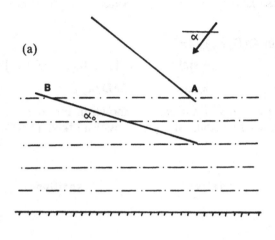

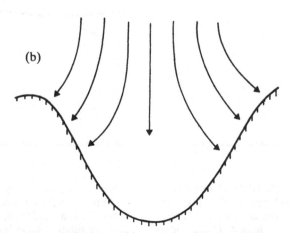

Fig. 6.28 Wave refraction. (a) Crest lines of oblique waves approaching a straight shore of uniform slope, dashed lines being underwater contours (see text for details). (b) Wave refraction in a bay.

$D/\lambda = 0.5$, the wave celerity starts to decrease. At some time later, point B on the same crest starts to feel the bed but point A has not travelled so far in the same time since the celerity is reduced. The result is that the orientation of the crests changes, the tendency being for the wave crests to coincide with the bed contours. The optical law of refraction applies, i.e. if α is the angle between wave crests and the shoreline in deep water and α_o is the equivalent angle in the shore zone, then the effect of water depth is such that

$$\sin \alpha / \sin \alpha_o = c/c_o = \lambda/\lambda_o \qquad (6.7.14)$$

One of the results of this refraction process is the pattern of wave action in bays—wave action is concentrated on headlands (Fig. 6.28b).

At the bed, motion beneath the waves is much simplified. The vertical movement disappears, resulting in a simple to and fro motion over a length, r_h, given by

$$r_h = H_o / \sinh 2\pi D / \lambda_o \qquad (6.7.15)$$

During the period T, a particle travels twice this distance so that the mean velocity, v_h, is given by

$$v_h = 2H_o / T \sinh 2\pi D / \lambda_o \qquad (6.7.16)$$

The period is unchanged—no waves are generated nor disappear so that $T = T_s$.

It is theoretically possible to combine knowledge of wave motion in the shore zone with criteria for sediment movement to produce analyses of bed stability similar to those for rivers (see, for example, Raudkivi, 1976) but the results are extremely complicated and unlikely to find application in ecology. Establishing the velocity at the bed and the approximate position of the breaker line for different wind, wave and water level conditions may provide an adequate account of the hydrodynamic regime in the shore zone. One result of interest is that relating the maximum velocity at the bed (eqn (6.7.16) with $2H_{so}$ replaced by πH_{so}) to the friction velocity (Lam and Jacquet, 1976). Since the friction velocity can be expressed in terms of stress, an indication of the conditions required to cause bed disturbance can be obtained.

6.8 OTHER FEATURES OF LAKE MOTION

6.8.1 Inflow and Outflow Dynamics

The relative importance of river flow in generating motion in a lake can be assessed by considering the rate at which energy is introduced into a lake by the flowing water. This can be compared to the rate at which energy is imparted by the wind (eqn (6.4.1)). For a river entering a lake with a mean velocity $\bar{U}$, and having a cross-sectional area A, the flow power, P_r, the rate at which kinetic energy is introduced, is given by

$$P_r = A\rho\bar{U}^3/2 \quad \text{(Watts)} \tag{6.8.1}$$

The relative energy inputs are illustrated in Table 6.3. The results are what might have been expected. Flood flows through small lakes subject to light winds are dominant while, with moderate winds blowing over large lakes, the influence of inflow is localised. The above definition of flow power must be distinguished from stream power which is the rate of potential energy expenditure per unit length of channel (Bagnold, 1966).

Where the density of the inflowing water is equal to or less than that of the lake surface water, river water remains on the surface and behaves like a form of jet. The boundary between this jet and the surrounding water is a surface of discontinuity. The result is instability, wave formation and the entrainment of lake water into the jet. The jet expands and decelerates until its velocity is indistinguishable from that of its surroundings. Most investigations of jet dynamics have been laboratory studies of jets discharging into still water and their applicability in lakes is uncertain. Figure 6.17 suggests that inflowing water can retain its identity for considerable distances and is influenced by the circulation in the lake.

There is no easy way of deducing the circulation pattern when there

Table 6.3   Relative energy inputs to a lake

(a) Flow power

		Flood flow	Intermediate	Low flow
Velocity, $\bar{U}$	m s^{-1}	2	1	0·5
Sectional area, A	m^2	50	25	10
Flow power	kW	200	12·5	1·25

(b) Wind power

		Light wind	Fresh wind	Storm
Wind speed	m s^{-1}	3	10	20
Wind power	kW km^{-2}	1·22	25·5	288

is high throughflow on the surface. In principle, eqn (6.5.4), which expresses the combined effect of drift and gradient currents, can be applied to the unequal transport balance. The difficulty is this tendency for the inflowing water to retain its identity over long distances, i.e. it is not certain that the throughflow, in moving through the lake, is uniformly distributed across it. The inflowing water can also induce secondary circulations in a somewhat similar way to that illustrated in Fig. 3.9.

If the river water is denser than the lake surface water, it will either sink to the bottom or to some depth where the lake water density is the same. Initially however, the momentum of the inflow causes it to remain on the surface until at some point, the plunge point, the forces due to the density difference are equal to those associated with the reduced momentum. Farell and Stefan (1988) suggest that the location of the plunge point is determined by water depth and not distance from the shore. They give the following equation for the depth at the plunge point, z_p,

$$z_p = a[q^2/(\Delta\rho/\rho)]^{1/3} \tag{6.8.2}$$

The coefficient, a, is between 1·3 and 1·6, $\Delta\rho$ is the density difference between river and lake surface water and q is the discharge per unit width of the inflow channel.

Inflows that plunge to the lake bed are often associated with high sediment loads rather than a result of temperature differences alone. Such turbidity currents flow like rivers under reduced gravity and may, initially, cause erosion of the lake bed. Again there is entrainment and mixing on the upper surface so that they ultimately lose their velocity and deposit sediment directly on the lake bed.

Alternatively, density currents may plunge to some depth in a stratified lake where the densities are equal, their fate depending on conditions in the lake. In the absence of internal wave motions, they are gradually weakened by entrainment but, like surface jets, they can retain their identity for long distances. Since there is no bed friction, the Coriolis force due to the earth's rotation may be significant, causing the density current to be deflected to the right in the Northern hemisphere. Depending on the value of the Wedderburn Number, there may be complex motion and mixing in the vicinity of the thermocline, causing rapid destruction of the density current.

The rate of outflow from a lake is controlled by the water level, i.e. by a stage discharge curve as in a river (Section 5.2.2). Knowing the discharge, Q, at a particular level, the velocity in the outflow channel, $\bar{U} = Q/A$ where A is the cross-sectional area of the channel. A length scale for the influence of this outflow velocity can be estimated by assuming that ideal fluid theory applies (Section 3.1.1). A series of concentric circles, equipotentials, are drawn, centred on the outlet. Streamlines are orthogonal to the equipotential lines, the principle being that the discharge through any volume bounded by streamlines is constant. More information on plotting streamlines is given by Francis (1958). Referring to Fig. 6.29, the velocity at A, for example, is given by the discharge between streamlines ($Q/8$ in this case) divided by an area equal to the distance between streamlines times the depth at that distance. By finding the distance where the outflow velocity calculated in this way is no greater than the velocity in the lake, the zone of outflow influence can be estimated.

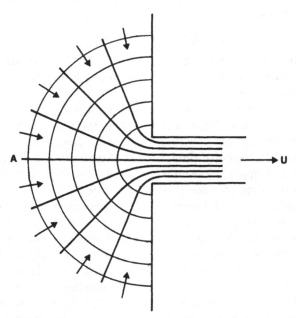

Fig. 6.29 Convergent flow around a lake outlet. The convergent circles are equipotentials and the orthogonal lines are streamlines. The discharge through any volume bounded by streamlines and the lake bed is constant, i.e., the discharge at A is one eighth of the total outflow.

6.8.2 Turbulence in Lakes

It is desirable to summarise the consequences of the various types and scales of motion in lakes in terms of turbulence coefficients. The terminology commonly used is not always consistent. Coefficients are variously referred to as the coefficient of eddy diffusion when, in fact, the mixing process involves dispersion arising from velocity gradients, eddy diffusivity and the coefficient of eddy viscosity. In addition, coefficients derived from analysis of temperature gradients, correctly relating to the estimation of eddy conductivity, are not always distinguished from coefficients derived from natural or introduced tracer studies. The value of such coefficients—referred to as diffusion coefficients, K, hereafter—is that they provide a measure of the overall mixing or turbulent intensity as implied in the notion of an equalisation time, T_e, equal to $(L')^2/K$ where L' is a characteristic length. Both Hansen (1978) and Imberger and Hamblin (1982) stress the uncertain effect of fetch—how large must a lake be for turbulence to be fully developed? All estimates of diffusion coefficients have large margins of error which is why the above inconsistencies are not important in practice.

Vertical turbulence
In the absence of stratification, equations for the vertical diffusion coefficient, K_z, have the same form as that derived from mixing length theory, i.e. eqn (3.2.4) can be written as

$$K_z = 6\cdot67 \times 10^{-2} U_f D \qquad (6.8.3)$$

Observations by Bengtsson (1973) in small Swedish lakes suggest that the coefficient is approximately 10^{-2}. The analysis by Banks (1975) of circulation in isothermal lakes gives the following equation

$$K_z = \tau_s D/4\rho U_s \qquad (6.8.4)$$

Since $U_f = \sqrt{\tau_s/\rho}$, this has a similar form, i.e.

$$K_z = (U_f/4U_s)U_f D \qquad (6.8.5)$$

Using the relations of Section 6.4.2, the various relations between K_z/D and wind speed are shown on Fig. 6.30. For smooth conditions

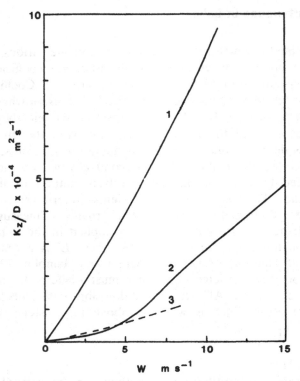

Fig. 6.30 The effect of wind speed, W, on the ratio of the vertical diffusion coefficient to water depth, K_z/D. Line 1 is derived from mixing length theory (eqn (6.8.3)). Line 2 is Banks' (1975) relation (eqn (6.8.5)). Line 3 represents the observations by Bengtsson (1973).

($W < 5\,\mathrm{m\,s^{-1}}$), the relations of Bengtsson and Banks are very similar, i.e.

$$K_z = 1 \cdot 2W \times 10^{-5}D \qquad (\mathrm{m\,s^{-2}}) \qquad (6.8.6)$$

A similar relation was confirmed in observations by DiGiano *et al.* (1978). Bengtsson's observations do not extend to winds $> 7\,\mathrm{m\,s^{-1}}$. Assuming that the mixing length relation is an upper limit which is not attained, Bank's relation is the best available estimate for winds $> 5\,\mathrm{m\,s^{-1}}$, i.e.

$$K_z = (4 \cdot 29W - 15 \cdot 45) \times 10^{-5}D \qquad (6.8.7)$$

The above equations can be used in the epilimnion of stably stratified lakes, replacing D by z_e, the thermocline depth.

In the presence of density gradients, mixing is much reduced. Lerman (1979), reviewing a number of investigations, shows that K_z is inversely proportional to the stability frequency (Section 6.6.2), i.e. $K_z \propto (N_*^2)^{-a}$. The value of a, however, is variable, ranging from 0·4 to 2·0, and the range of plotted values covers four orders of magnitude $(10^{-6}–10^{-2})$, i.e. from the molecular value to the epilimnion values implied by eqn (6.8.6). Ward (1977) attempted to summarise the available data, incorporating the lake area, A_L, i.e.,

$$K_z = 4·9 \times 10^{-12} g \sqrt{A_L}/N_* \tag{6.8.8}$$

The search for simple relations describing vertical diffusion in the presence of density gradients is inherently unrewarding. Density structure, though important, is not the only factor derermining the intensity of turbulence.

Horizontal turbulence

The one dimensional advection–diffusion equation (Section 3.2.4) shows that introduced tracer is, subsequently, normally distributed. It is easy to envisage that, in two dimensions in the absence of advection, the resulting distribution is a series of concentric circles of tracer concentration contours such that the cross-section on any diameter is normally distributed. In the presence of advection, the pattern becomes a series of ellipses (Fig. 6.31). A measure of the spread of tracer in the elliptical configuration is given by the variance, σ, where $\sigma = 2\sigma_x\sigma_y$, σ_x and σ_y being the standard deviations in the x and y

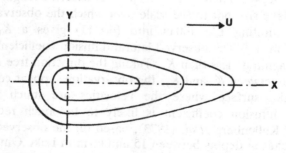

Fig. 6.31 Contours of tracer concentration on a lake surface in the presence of advection.

directions. An equivalent coefficient for two dimensional diffusion, K_h, is usually taken as

$$K_h = \sigma^2/4t \qquad (6.8.9)$$

The notion of an eddy spectrum (Section 3.2.1) emphasises that, as floats or tracer particles separate, the likelihood of them being pulled further apart by larger eddies increases. The equivalent diffusion coefficient, therefore, increases with the elapsed time or distance from the origin. Various relations between K_h and a characteristic length, L', have been proposed, e.g.

(Okubo, 1971) $K_h = 1\cdot99 \times 10^{-4}(L')^{1\cdot15}$ (6.8.10)

(Lam and Simons, 1976) $K_h = 2\cdot23 \times 10^{-4}(L')^{1\cdot3}$ (6.8.11)

These can only indicate average values since one would expect the spread of surface tracer to be directly affected by current hydrodynamic conditions. Leenen (1982) suggests that

$$K_h = U_f L' \qquad (6.8.12)$$

The various relations are compared in Fig. 6.32. Leenen (1982) also suggests that the characteristic length is not the dimension of a lake but the size of the largest eddy. This is determined by the overall circulation pattern and may vary with wind direction within the same lake.

Bengtsson (1973) also investigated horizontal dispersion using floats. An unexpected feature is that the coefficient in the direction of the surface, K_x, is virtually the same in different lakes at different wind speeds, viz. $K_x = 2\cdot5 \times 10^{-2}\,\mathrm{m\,s^{-2}}$. The characteristic length here is not related to lake size but to the scale over which the observations were made. Substituting $L' = 100\,\mathrm{m}$ into (6.8.11) gives a K_h value of $8\cdot87 \times 10^{-2}\,\mathrm{m^2\,s^{-1}}$. The observed lateral diffusion coefficient, K_y, is an order of magnitude less than K_x. Taking the direction free K_h value as somewhere between K_x and K_y, the observations appear consistent.

Below the surface where the velocities are much lower, the horizontal diffusion coefficient is likely to be much reduced. The formula of Kullenberg *et al.* (1973), based on the observed spreading of dye patches at depths between 15 and 60 m in Lake Ontario, is

$$K_h = 1\cdot1 \times 10^{-8}(L')^{1\cdot5} \qquad (6.8.13)$$

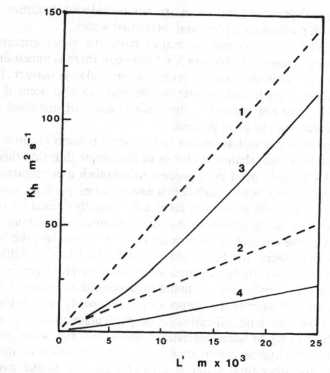

Fig. 6.32 Variation of the horizontal diffusion coefficient, K_h, with length scale, L'. The dashed lines are Leneen's (1982) relation (eqn (6.8.12)), line 1 being for a wind speed of $5\ m\ s^{-1}$ and line 2 for $2\ m\ s^{-1}$. Line 3 represents Lam and Simon's (1976) relation (eqn (6.8.11)) and line 4 Okubo's (1971) relation (eqn (6.8.10)).

6.8.3 Exchanges between Sediment and Water

This section considers the physical processes controlling the flux (mass per unit area and time) of solid and dissolved material in and out of lake sediments. Sediments have a key role in lake ecology—are they a sink in which material is trapped forever or a temporary store in a recycling process? Where materials such as nutrients are recycled, lakes exhibit some of the features of a closed system in that dependence on external sources is reduced. With reactive substances, the quantities transferred are controlled by chemical reactions as well as by physical processes but these are not considered. Sediment here implies both particulate matter and interstitial water. The deposition of particulate matter from the water column has already been discussed in Section 3.4.3 although there is uncertainty as to whether the assumption of no resuspension is always correct. Direct accumulation of particulate matter on the bed can also occur if there are turbidity currents present. The incorporation of dissolved solids into sediment is a chemical process.

The return of particulate matter to the water column requires some form of physical disturbance. In terms of the simple diffusion theory of Section 3.4.3, there must be a process to establish a concentration at the reference level before turbulent mixing carries particles upwards. The largest hydraulic stresses in lakes are normally caused by surface waves so that wave action is the most common disturbing force. Referring to the circulation pattern on Fig. 6.17, however, the bottom return current along the South side of St Serf's Island is sufficiently strong for the bed to be an area of erosion in Hakanson's (1982) terminology (see below)—any material deposited is removed by the current alone. Whatever the nature of the sediment, i.e. whether it is firm or a viscous fluid, disturbance of particulate matter is accompanied by mixing of some interstitial water into the water column. Even without physical disturbance, if the concentration of dissolved matter in the interstitial water is greater than that in the overlying water, material will be transferred by diffusion, the amount depending on the diffusion coefficient and the concentration gradient.

The role of sediments in the ecology of lakes would be clearer if the mass balance of Section 4.1.3 could be extended to include a balance equation for the sediments. This has been done by Lorenzen (1974) but his assumption of constant rates or velocities for the flow of matter in and out of the sediment is very oversimplified. The processes involved are neither spatially uniform nor constant in time.

Spatial variation in sediment–water exchange is examined by Hakanson (1982). From a study of sediment characteristics in Swedish lakes, Hakanson identified three bed zones which are analogous to the traditional, longitudinal division of a river, viz.

—areas of accumulation where fine material having high organic and water content is deposited;
—areas of transportation where quite long periods of accumulation are interspersed by bursts of disturbance and transportation;

—areas of erosion where there is no deposition of fine material, the bed consisting of rock, gravel or glacial clay.

Hakanson's analysis considers how the relative areas of these bed zone types vary with lake morphometry and, in particular, how the combined areas of transportation and erosion vary with the dynamic ratio—the square root of the lake area divided by the mean depth. When the ratio exceeds 3·8, the combined areas are equal to the total lake area, the whole lake being dominated by wave action. At the other extreme, the combined areas never fall below 15 per cent of the total but, with the steep slopes associated with low values of the dynamic ratio, the effects of slope are more important than wave action. Deposited sediments are not stable on slopes greater than 14 per cent, i.e. the disturbing force is gravity. McManus and Duck (1983) describe various subaqueous landforms observed in sonar scans that are believed to be caused by the slumping of unconsolidated sediments.

Hilton (1985) seeks to explain the causes of sediment focussing—the greater accumulation of sediment in parts of a lake—primarily from an analysis of trap data. The processes identified as the causes of redistribution broadly coincide with those of Hakanson (1982) but the trap data reveal the disturbance associated with the overturn of a stratified lake as a further cause. Again it is shown that it is the relation between lake area and depth that determines the controlling processes.

A number of investigations demonstrate that, in lakes where the areas of transportation and erosion form a significant part of the total lake area, quantitative relations between predicted wave characteristics and the quantity of matter returned to the water column can be established. Many of these are directed towards assessing the role of resuspension in the recycling of phosphorus. A key study is that of Lam and Jacquet (1976) who show that, in Lake Erie, the amount of regenerated phosphorus can be related to the velocity of the motion beneath the waves and the sediment grain size. A useful presentation is that of Carper and Bachmann (1984) who, effectively, introduce the time element into Hakanson's bed zone types by plotting the fraction of the bed area affected by waves, the transport and erosion areas, against the percentage time. Aalderink *et al.* (1984), reviewing a number of formulae, show that reasonable predictions of suspended sediment concentration—the net result of deposition and resuspension—are possible.

6.9 FINAL REMARKS

The preceding sections outline the main features of the hydraulic structure of lakes. Table 6.4 is an oversimplified summary, the distinction between vertical and horizontal structure being somewhat arbitrary. Vertical structure emphasises features that extend over the whole lake although their dimensions may vary spatially while horizontal structure emphasises features having a definite location. That lake hydrodynamics are, inherently, complex and not fully understood is obvious, the difficulties being compounded by the lack of observational data.

To be set against this complexity is the limit to the amount of detail

Table 6.4 Summary of hydraulic structure

Structure	Associated motion
(a) Vertical structure	
Surface mixed layer	Orbital motion beneath waves
	Langmuir circulations
(i) Isothermal conditions	
Ekman layer	General circulation
	Surface seiches
Bottom boundary layer	
(ii) Stratified conditions	
Epilimnion	General circulation
	Surface seiches
Thermocline	Kelvin–Helmholtz instability
	Density currents
Hypolimnion	Internal seiches
	Turbulent patches
	Turbidity currents
Bottom boundary layer	
(b) Horizontal structure	
General circulation	
Upwelling/downwelling zones	Vertical motion
Principal circulation zones	Horizontal motion
Secondary circulation zones	Induced rotations
	Motion in partially enclosed bays
Other horizontal differentiation	
Inflow zone	Surface jets
Outflow zone	Convergent flow towards outlet
Shore zone	Wave action dominant
Edge effects	Thermocline/lake bed

that can be incorporated into ecological investigations. It is usually only short term experiments that involve biological sampling at less than daily intervals and, in many cases, the sampling interval is much longer. There are similar restrictions on the amount of spatial detail. Application of hydrodynamics, therefore, usually involves some form of averaging in both time and space. But this is not an argument for oversimplifying hydrodynamics. The explanation of observed biological features may lie in the details and in short term events. For example, the rate of release of dissolved material out of lake sediments may be determined by residual motion from earlier oscillations. It is important, therefore, to know where such details of the motion are likely to be important, i.e. where its direct observation is essential and where average or predicted values are adequate.

Although the two are related, it is convenient, when considering what investigations are appropriate, to view a lake as a chemical reactor separately from an examination of lake habitats. The reactor view of a lake stresses its mixing characteristics, e.g. its possible compartment model structure, resulting residence time distribution and the role of sediments in recycling material within it. The notion of hydraulic structure originated by analogy with the role of air masses in meteorology. In terms of habitat, however, all it does is add more detail to the classic limnological division of a lake into pelagic (open water) and benthic (boundary layer) habitats. The gain is more precise delineation of the relation between habitat characteristics, lake morphology and the surrounding climate.

7

River Basin Modification

Few rivers are in their natural, pristine state. For the majority, the structure and dynamics described earlier have to be modified to take account of the changes that have taken place. Attention is focussed on the changes that directly affect the relation between water movement and freshwater ecology. The extensive literature on the impact of development is not reviewed. The areas of particular interest are:

—alterations to natural flows resulting from reservoir operation, abstractions and discharges;
—the differences between reservoirs and lakes modified to regulate flow and natural lakes;
—alterations to river habitats caused either by changes in river morphology as a result of engineering work or the effect of flow regulation.

Other changes are, ultimately, reflected in the above. Changes in land use within a catchment, for example, may alter the volume of runoff entering the river system and its distribution in time. Even possible effects of atmospheric pollution and climate change must, eventually, be examined in terms of hydroclimate although the impact may be much wider, extending, for example, to the wind regime and the thermal characteristics of lakes. This, however, is not considered here.

7.1 RESERVOIRS AND FLOW REGULATION

7.1.1 Types of Reservoir

The feature of natural river flow is its variability and, without storage, the guaranteed flow or reliable yield is the lowest flow. Major engineering works in river basins, therefore, usually involve regulation of the natural flow through the construction of reservoirs. In most developments—water supply, hydro-electric power generation and so on—water diverted as a result of reservoir operation appears somewhere else although it may be in a sewage outfall in a different catchment. Only in the case of irrigation is water totally lost through evaporation.

In terms of operation, on-river reservoirs are of three basic types;

—direct supply reservoirs where the supply is abstracted directly: compensation water may be discharged into the downstream river;
—regulating reservoirs where the object is to control the flow in the downstream river for navigation, for example, or to abstract water from the regulated river;
—flood control reservoirs where the object is simply to reduce the risk of flooding downstream (see Fig. 7.1).

In addition to on-river reservoirs, abstraction reservoirs may be constructed adjacent to a river, particularly in the lower reaches. Although their inflow is regulated, the permitted rate of abstraction usually depends on the current state of the river.

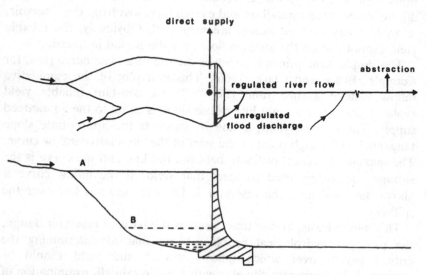

Fig. 7.1 Schematic reservoir layout. Level A represents maximum capacity and line B the minimum effective capacity.

In practice, many of these features are combined. A single reservoir may have multiple objectives—to supply water,. to regulate flow downstream, to prevent flooding and to provide recreational facilities. Often these objectives conflict. Flood control is easier if the reservoir is kept empty, for example, so that quite complex rules for optimum operation have to be defined. Also, a particular reservoir may form part of a linked system. A city's water supply may be drawn from a number of reservoirs, river intakes and underground sources. Again, there are likely to be optimisation rules controlling the overall system. Hydro-electric power schemes often involve a linked series of reservoirs so that the flow downstream of a scheme does not reflect irregular power demands. The pattern of flow in affected rivers, therefore, may bear little relation to the natural flow and direct measurement of discharge is necessary.

7.1.2   Storage Theory

The question to be considered is, given the pattern of inflow over a period, what storage volume is required in order to give a constant, guaranteed supply over that period. A secondary question, of considerable ecological interest, concerns the resulting water level fluctuations. For simplicity, the corrections necessary to account for the difference between rainfall on and evaporation loss from the reservoir, as well as any seepage losses, are neglected. Obviously, the reliable yield cannot exceed the average flow over the period in question.

The simple basic principle involves the use of a mass curve (see, for example, Bruce and Clark, 1966). This is a plot of the cumulative inflow volume against time (Fig. 7.2). A constant reliable yield (volume/time) is a straight line whose slope is equal to the guaranteed supply. For a given yield, a line is drawn at the appropriate slope tangential to the high point at the start of the cumulative inflow curve. The maximum vertical ordinate between the line and the curve is the storage volume required to meet that yield. If the inflow curve is above the yield line, the reservoir is full and water is lost over the spillway.

This method was, at one time, used as the basis of reservoir design, the principal hydrological problem being that of determining the critical period over which inflow, storage and yield should be compared. In temperate climates with reliable rainfall, examination of reservoir response over the three consecutive driest years is usually

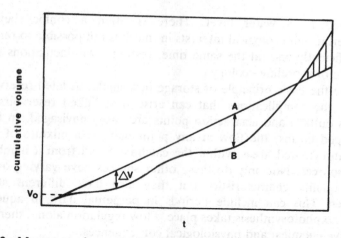

Fig. 7.2 Mass curve for estimating the reservoir storage required to meet a specified yield. Line A represents the yield and line B the cumulative inflow. AB is the storage volume required and the hatched area represents spillage loss. At any time, ΔV is the change in volume compared to the initial value, V_o. By making use of a volume–depth curve for the site (Section 6.1.2), this information can be used to predict level fluctuations.

considered sufficient but, in drier areas with less reliable rainfall, periods up to 10 years may be required.

Actual reservoir design obviously requires extension of the basic method. Besides the corrections for rainfall and evaporation mentioned above, demand is likely to vary seasonally and economic considerations require the investigation of the relation between different storage volumes and potential yield. In addition, there are other engineering and operational requirements. For example, there is likely to be a minimum water level below which water cannot be extracted, quite apart from the fact that abstraction should not disturb the bed and create a host of water quality problems. Modern design methods involve numerical simulation of reservoir behaviour over long periods so that relations between flow, storage and the risk of failing to meet a specified yield as well as operating rules for multi-purpose reservoirs can be developed.

If the actual reservoir site is known, i.e. a volume–depth curve exists, then the pattern of water level fluctuations can be determined. If the reservoir volume at the start is V_o, then the vertical ordinate on Fig. 7.2 at any subsequent time is the volume drawn from storage, ΔV, and the current reservoir volume is $V_o - \Delta V$ which can then be converted to a water level. There is often a conflict between engineering and ecological interests in that it is not possible to regulate flow efficiently and, at the same time, restrict level fluctuations in the interests of shoreline ecology.

While the basic principle of storage in a single, isolated reservoir is simple, the complications that can arise with linked reservoirs with various outfalls and abstraction points are easily envisaged. In highly developed basins, the flow at any point may be a mixture of water from unregulated areas within the catchment and from a number of other sources. Not only do these other sources have varying original water quality characteristics but they also have different storage histories. This can include periods in perpetual dark in aqueducts where no photosynthesis takes place. Flow regulation alone, therefore, can have chemical and physiological consequences.

7.1.3 Reservoir Features

Reservoirs are subject to the same external factors, sun, wind and throughflow, as natural lakes. The majority of the features of the hydrodynamics of lakes apply to reservoirs and they have similar hydraulic structure and habitats. There are, however, some differences which influence the ecology of reservoirs.

In a natural lake, the outflow rate is dependent on the current water level but the basic principle of reservoir operation—balancing the inflow fluctuations—destroys this element of self-regulation. Water levels are drawn down much lower so that the chracteristic feature of many reservoirs is the scar round the shore, the absence of vegetation and the general paucity of species. In hydro-electric reservoirs in particular, the level fluctuations can be rapid and irregular.

Reservoirs have a characteristic morphology. In longitudinal section, they appear, in comparison to a natural lake, to be truncated by the dam. In itself, this appears to have no obvious effect and may make the assumption of a rectangular basin more realistic. In plan, the result of inundation may be number of side arms in what were the tributaries of the main river. In these side arms, the motion induced by inflowing water may be more important than that due to wind while the opposite is true of the core of the reservoir and there may be other effects such as additional oscillations in these branches. As a result, there may be greater spatial variation in reservoir habitats.

Water is usually withdrawn from a reservoir at some depth below the surface. This is of little ecological consequence in isothermal conditions although it may result in the distortion of the vertical temperature profile if the reservoir is stratified. The hydrodynamics of withdrawing water at depth in stratified conditions are complicated and reviewed by Imberger (1980). Problems can also arise as a result of thermocline tilt and internal seiching. For a small value of the Wedderburn Number, the tilt and seiche amplitude can be appreciable and may reach the level of the draw-off points. This is a problem for the reservoir operator if de-oxygenated, hypolimnetic water enters the supply system but it is an ecological one if this water is discharged into the downstream river.

7.2 MODIFICATIONS TO RIVER HABITATS

7.2.1 Introduction

River habitats can be altered by changes in flow regime and by changes in the structure of the river itself as a result of engineering work. River reaches may be affected by both forms of change and the two may interact, i.e. changes in flow regime may alter channel geometry. Effectively, the dominant discharge is altered and the river seeks to adjust its dimensions to the new conditions. Although the presence of a reservoir may reduce the magnitude of the more frequent floods, and hence the dominant discharge, it may not reduce the severity of extreme floods. Because the channel dimensions have been altered, the consequences of extreme floods may be more serious. Active rivers are likely to adjust more rapidly than stable ones.

Additional effects are caused by the presence of the reservoir itself. Besides chemical changes resulting from the increased retention time and possible changes in water temperature, the downstream river has a steady source of plankton and reduction of the solid load may affect river morphology since the sediment mass balance is altered. The dam is a barrier, particularly to fish movement unless some form of fish pass is installed.

7.2.2 Effects of Flow Changes

It is usually assumed that changes in flow regime imply flow reduction but this is not necessarily so. If the flow is reduced, the immediate changes in river habitats are:

—reduction in the actual habitat area;
—change in the nature of the boundary layer flow, i.e. in the value of the relative roughness and, to a lesser extent, in slope;
—alterations to the sequence of hydrodynamic conditions that occur and in the resulting downstream transport of material.

It is unlikely, even in regulated conditions, that the flow is absolutely constant. Correctly, therefore, the first two of the above are expressed as changes in the frequency of occurrence of different states of flow. Change in the sequence of hydrodynamic conditions emphasises that affects may not be limited to the immediate area of change. Changes in stream drift and the food supply of downstream species may result.

An important question concerns the desirable features of a regulated flow. There is an inherent conflict here. To the operator of a water utility, compensation flow is a loss—a proportion of the cost of providing storage is not reflected in available water. It is difficult to give direct answers. The problem is, partly, biological, since the relation between hydrodynamic conditions and biological success, however measured, is neither direct nor easily demonstrated. Many factors are involved and biological conditions may not be the only criteria for determining compensation flow. Recreational use and aesthetics may be just as important and, indeed, may contribute towards the cost of providing compensation water.

The simplest case of the compensation flow below a direct supply reservoir is considered. Traditionally, compensation flow is fixed as a specified fraction of the average flow, i.e. as the flow exceeded x per cent of the time and with little or no reference to downstream conditions. It is relevant to consider the length of a river in which compensation water is a major part of the flow, i.e. above the confluence of major, unregulated tributaries. This length can be compared to the overall river network. If the affected length is y km of a third order stream, what proportion of the total network does this represent?

More recently, recognising that occasional high flows are beneficial, block releases have been introduced. Here an annual volume of water is agreed as the compensation, to be released downstream when this is considered most desirable. Where constant flow is specified, then, for *x* per cent of the time, the compensation flow is greater than the natural flow. This may be considered wasteful and it may be better to attempt to reproduce some of the features of the natural seasonal cycle. More usually, block releases are intended to reproduce spates. To be worthwhile, therefore, the change in flow must be sufficiently great for the downstream hydrodynamic conditions to be altered. For example, at least some disturbance and transport of deposited organic material should occur. The consequences of a sudden release of water into a river having a low initial flow are described by Petts *et al.* (1985). The release is large and perhaps more typical of the sudden operation of a hydro-electric plant but the modification of the wave form as it travels downstream (Fig. 7.3) as well as changes in water quality, particularly suspended solids concentration, are demonstrated. The wave form is quite rapidly modified and, although the wave may be detected a considerable distance downstream, the benefits, in terms of modified hydrodynamic conditions, may be restricted to relatively short distances.

7.2.3 Changes in River Structure

Few rivers, particularly in their lower reaches, have retained their natural morphology and dimensions. Some effects are localised such as bank protection works near roads and bridges which restrict natural changes in alignment, but major alterations to river habitats can result from attempts to improve land drainage and reduce flooding. Increasing the capacity of a channel to discharge water can involve changes in the slope and alignment of the channel as well as in its sectional form and roughness. The latter may involve the cutting and removal of vegetation as well as the construction of flood banks. If the channel is made too large, it will attempt to adjust to the dimensions appropriate to the flow regime, resulting in the build-up of sand and gravel bars within the artificial channel. The use of a river for navigation requires the maintenance of a minimum water depth irrespective of the discharge, restrictions on current speed and thus on slope and, at times, changes in channel alignment. To achieve these conditions, dredging and the construction of locks may be necessary. River banks are often protected since the passage of boats generates waves that

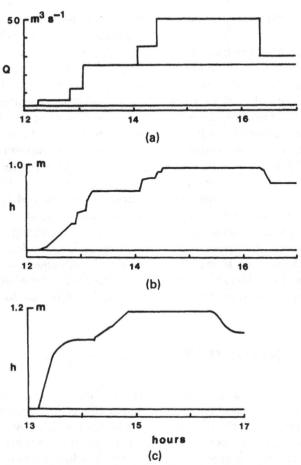

Fig. 7.3 Changes in the wave form in a river during the downstream passage of a sudden release from a dam (based on Petts *et al.*, 1985). (a) Original release. (b)–(f) indicate recorded water levels at various distances from the release point. (b) 0·25 km; (c) 2·0 km; (d) 4·35 km; (e) 10·05 km; (f) 14·88 km.

cause erosion as well as additional turbulence that may disturb the bed.

The catalogue of activities that have changed river habitats could be expanded at length. Some of the changes occurred so far back in the past that they are probably, in the public perception, an intrinsic part of the river. Some old mill weirs are an obvious example. There is increasing recognition, however, that the more recent drastic changes

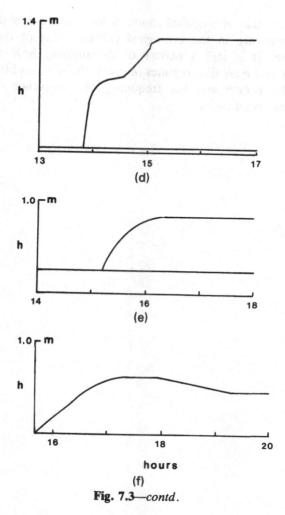

Fig. 7.3—*contd.*

associated with land drainage and flood control are not desirable. This realisation reflects not only ecological concern about loss of habitat and associated species but also sound river management. Confining water within the river channel can have serious consequences downstream. The natural process of self-regulation within a river basin involves flood water being stored temporarily on the flood plain. Careful design of the river, involving the same principles as outlined in Section 5.1.3 can improve the discharge capacity and retain viable river habitats.

The examination of modified channels involves no new principles— the differences are in the ecological consequences of the modified hydroclimate. It is still a matter of determining how the altered morphology and river flow regimes interact, the nature of the resulting boundary layer flow and the frequency of occurrence of different hydrodynamic conditions.

8

Freshwater Ecosystems

8.1 IDEALISED ECOSYSTEMS

8.1.1 Ecosystems and Ecosystem Models

The original notion of an ecosystem (Tansley, 1935) emphasised the idea of location—'. . . not only the plants and animals of which it is composed but the animals habitually associated with them, and also all the physical and chemical components of the immediate environment or habitat which together form a recognisable self-contained unity'. Modern conceptions emphasise the interaction between living and non-living parts and the flow of energy and materials between these parts. This leads to the diagrammatic view of an ecosystem shown in Fig. 8.1. The essential features of this dynamic view of ecosystems are: interactions between plants and animals; an external energy supply; a source of raw materials, partly external and partly recycled by decomposition within the ecosystem; some losses from the system; processes regulated by the external environment. In addition, the ecosystem has a characteristic habitat structure created by the combined effects of the biotic and abiotic features. Some confusion can arise due to the different interpretations of environment. In systems dynamics, environment refers to all those characteristics which are external to the system being investigated whereas, in ecology, environment includes the habitat characteristics which are within the system being investigated. One of the major difficulties in applying the ecosystem concept is knowing where to draw the boundaries, what actually constitutes an ecosystem.

The core of most ecosystem models is the analysis of changes in time in some measure of the populations (biomass, numbers, units of

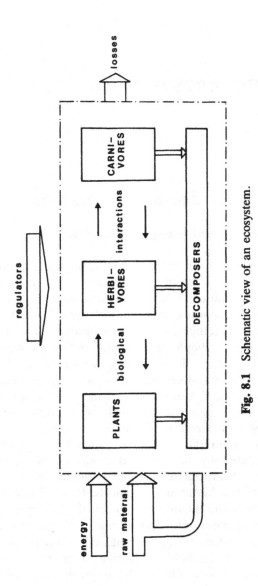

Fig. 8.1 Schematic view of an ecosystem.

energy) of the different trophic groups within the system. Secondary investigations, such as the relative importance of the recycled and externally supplied raw materials may be almost as important.

Since the rate of change of biomass (growth or decay) is dependent on the present biomass, B, state or growth equations have the following essential form

$$dB/dt = rB \qquad (8.1.1)$$

where r is a variable rate which takes into account time variations in the raw material and energy supply, environmental conditions and the biological interactions between species, i.e. it expresses the net effect of all factors causing both increase and decrease in population. This appears to be mathematical obscurantism. The justification is that it is acceptable to separate the effects of biological interactions from those of the physical conditions. Biological interactions, the consequences of the interactions between different trophic levels, can be summarised, very briefly, in terms of the trophic level at which control on population numbers is exercised:

—control from below   available food supply from prey
—control at same level   competition
                         breeding success
                         territorial behaviour
                         migration
—control from above   predation
                         parasites
                         disease

The starting point for analysing the effect of physical conditions, therefore, is through their influence on the variable growth rate, r. Separating out the influence of biological interactions is, clearly, open to criticism—such interactions do influence the impact of physical conditions. But overemphasis on the inter-connections and feedback effects can also inhibit analysis.

If the growth rate, r, is constant and positive, the solution to eqn (8.1.1) is exponential or Malthusian growth, i.e.

$$B = B_0 e^{rt} \qquad (8.1.2)$$

where B_0 is the initial population. More generally, where r varies in time, the solution (Murdie, 1976) is

$$B = B_0 \int e^{r(t)\,dt} \qquad (8.1.3)$$

The growth rate, $r(t)$, may at times be negative and cause the population to fall. An analytical solution is only possible if the expression for the time variation in growth rate, $r(t)$, can be integrated. Where $r(t)$ expresses the net effect of many different processes, this is not possible and a numerical procedure must be used (see Chapter 9).

The form of growth or decay described by eqn (8.1.1) can be termed rate controlled, i.e. growth is continuous in the mathematical sense that there are no sudden discontinuities and the population rises and falls solely in response to the seasonal and other changes in time of the growth rate (see Fig. 8.2). Intense storms and other events can cause virtually instantaneous falls in population. In a sense, this is a limiting case of rate control where the loss or death rate is infinite but it is better considered separately as event control. A third case arises when the population change depends not only on the growth rate but also on some upper limit to the population, the carrying capacity. The simplest possible case, where the growth rate is constant but where the population change is also proportional to the difference between the current population and the carrying capacity, gives rise to the logistic equation (see, for example, Pielou, 1969). Neither the growth rate nor the carrying capacity need be constant and, in general, this type of growth may be termed space controlled.

These three types of growth form the basis for a systematic

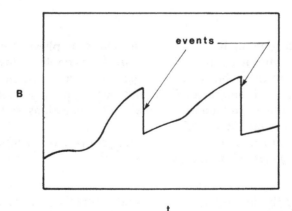

Fig. 8.2 The effect of rate and event control on biomass change. Rate control causes continuous change while events cause virtually instantaneous falls in biomass.

investigation of the physical conditions but, before going any further, it is sensible to convert the abstractions of Fig. 8.1 into views of rivers and lakes that might be recognised by someone standing at the water's edge.

8.1.2 Idealised River Ecosystems

The obvious feature of a river ecosystem is its linear structure, dominated by the uni-directional downstream flow. Processes at a particular point on a river are clearly dependent on what happens upstream and there is no obvious, *a priori*, basis for subdividing a river into component ecosystems.

Earlier river investigations stress the longitudinal variation in habitat characteristics and their associated plant and animal species, practical issues such as fishery management or the use of benthic invertebrates as indicators of water quality. More recently, Vannote *et al.* (1980) attempted to rectify this absence of a coherent theoretical basis for river ecology by defining the River Continuum Concept. This has proved somewhat controversial, much of the criticism being summarised by Statzner and Higler (1985). At the same time, the original promoters of the concept enunciated what they saw as the developments in river ecosystem theory (Minshall *et al.*, 1985). These arguments are not repeated here in any detail, the intention being to see what picture or model of a river ecosystem emerges from the controversy and, in particular, the part played by physical conditions in river ecology.

At the heart of the controversy is the admirable attempt to link river biology and ecology to the physical structure and dynamics of a river—'biological communities should become established which approach equilibrium with the dynamic physical conditions of the river' (Vannote *et al.*, 1980). Starting from this, a series of arguments and hypotheses lead to a continuously changing pattern of community structure and dynamics in a downstream direction, the latter expressed as stream order. This, together with subsequent modifications, is examined further below. Statzner and Highler's comments which are particularly relevant to the part played by the physical conditions are:

—that stream order is not a meaningful description of the physical environment;
—that the characteristics of the physical environment are determined primarily by water depth and slope and that these change discontinuously at a variety of length scales;

—that catastrophic events, such as floods, can cause major changes in species type and number and that it is legitimate to think in terms of succession in a river in the same way as for terrestrial ecosystems.

The more recent developments in stream ecosystem theory emphasised by Minshall *et al.* (1985), besides the apparent abandonment of the controversial and partly unproven analogies between the physical and biological energy distribution in a river, are all useful in clarifying the nature of river ecosystems. The most valuable contribution is, perhaps, the controversy itself and the resulting attempt to move stream ecology away from its descriptive phase to the formulation of hypotheses about the ecology of whole rivers that can be tested.

The key hypothesis of the river continuum concept is orderly change in the structure and function of plant and animal communities which reflect the changing physical environment. The assumption that the physical features are described by stream order has already been commented on. The concept also includes hypotheses about biological interactions within a river, stressing that community structure is not determined solely by physical factors. These are not considered here. The general structure of river communities is outlined in Fig. 8.3.

Central to the concept is variation in the source of the primary food supply, the lowest trophic level. In the upper reaches, the system is driven by allocthonous material, i.e. organic matter derived from the catchment. In the middle reaches, photosynthesis becomes increasingly important but this is inhibited in the lower reaches by increasing turbidity and the primary food source is organic matter derived from upstream although both phyto- and zooplankton do occur. The more recent modifications of the original scheme follow Rzoska (1978) in placing more emphasis on the nature of the catchment in determining the primary food supply. More recently still, Reynolds (1988) has demonstrated that phytoplankton have a more important role in the upper and middle reaches than was believed previously.

Associated with this variation in primary food supply are changing communities of bottom dwelling invertebrates, functional groups classified in terms of their feeding behaviour. The main groups and their food sources are as follows:

—shredders   coarse particulate matter derived from land;
—collectors   fine particulate matter filtered out of flowing water or gathered up from sediments;

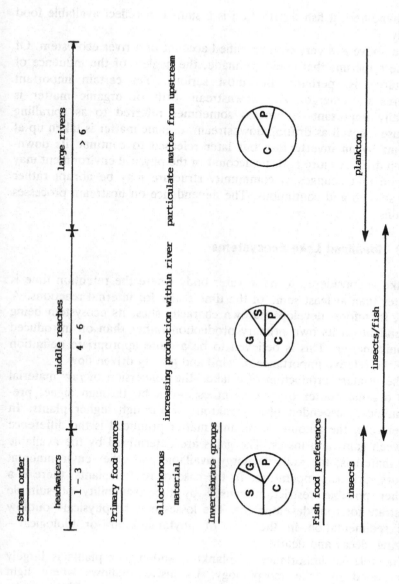

Fig. 8.3 Features of the idealised ecosystem based on the River Continuum concept. Numbers at the top refer to the assumed stream order number. S shredders; C collectors; G grazers; P predators.

—grazers   mainly algae attached to sediment surfaces;
—predators   other invertebrates.

The hypothetical fish distribution is assumed to reflect available food supply.

The above is a very oversimplified account of a river ecosystem. Of all the criticisms that could be made, the neglect of the influence of tributaries is, perhaps, the most serious. Yet certain important features do emerge. The downstream drift of organic matter is crucially important—a process sometimes referred to as spiralling because, as well as drifting downstream, organic matter is taken up at a point by an invertebrate but later released to continue its downstream drift. A more realistic account of the physical environment may indicate that changes in community structure may be abrupt rather than smooth and continuous. The dependence on upstream processes remains.

8.1.3 Idealised Lake Ecosystems

A lake is considered to be a water body where the retention time is greater than at least some of the time scales for internal reactions. A lake, therefore, develops its own characteristics, its ecosystem being dependent on its own primary production rather than on introduced organic matter. This is believed to be a more appropriate definition than the relative importance of wind and gravity driven flow.

The primary production of a lake—the conversion of raw material into organic matter by photosynthesis—can be through algae, predominantly suspended phytoplankton, or through higher plants. In either case, the amount of organic matter produced is the difference between gains and losses. The gains are determined by the available raw materials, the radiant energy available and other environmental factors such as temperature. In the case of rooted plants, there is a further possible restriction caused by the availability of suitable substrate for root development. The losses may be physical—outflow and sedimentation in the case of phytoplankton—or biological—grazing, decay and death.

The relative importance of plankton and higher plants is largely determined by lake morphology. Extensive shallows where light penetrates to the bed and where the substrate is suitable, may be completely dominated by higher plants whereas, in a deep, steep sided lake, planktonic production is all important. Inevitably, there are exceptions. Floating vegetation, particularly in tropical lakes, may be a source of primary production and, in some shallow waters, very high algal densities may so reduce light penetration that no rooted plants survive. In some small water bodies, as in rivers, external sources of organic matter may be important.

In comparison to terrestrial and marine environments, lake are, in many ways, ideal ecosystems for study. Their boundaries are clearly defined and most of the inputs and outputs are via the river system. Yet complete investigations of the links implied in Fig. 8.1 are comparatively rare. Most investigations focus on particular aspects of lake ecology and the overall interdependence is not revealed. Partly this is because of the complex food webs that occur—the structure of Fig. 8.1 is very oversimplified—but, in addition, many of the practical issues concern either water quality or fisheries where the requirements are rather different.

Correctly, the hydraulic structure must be known before the organisation of the ecosystem can be described. Yet spatial variation is often neglected. In investigations of water quality where the prime concern is undesirable plankton densities, the whole lake is reduced to a single representative column and only vertical variations within it are considered. Horizontal variation is reduced to a sampling problem—how to obtain lake mean values of population density, chemical and physical characteristics. Other features of the hydraulic structure may be taken into account, e.g. in relation to the performance of a lake as a chemical reactor where the residence time distribution or the recycling of nutrients may have to be considered.

Fisheries investigations start at the other end of the food chain and consider recruitment, available food, the size and quality of fish and, above all, the sustainable yield—the number of fish that can be removed without depleting the population. Little importance is attached to the dynamic features of the hydraulic structure, interest in spatial variation being focussed on more static features such as the extent of feeding grounds and spawning beds. The stages between water quality and fisheries investigations—between the grazing pressure on phytoplankton and the availability of food for harvested fish—is often neglected.

The relative lack of interest in intermediate stages of the food chain reflects management concerns and the availability of scientific resources. This, together with the neglect of hydraulic structure, can lead to failure to detect interdependence within a lake ecosystem. It is not possible to specify a characteristic lake ecosystem for there is too much diversity in the biological interactions as well as in the physical environment. The choice of what is included in any view of an ecosystem is determined by the interests involved. Such simplifications are acceptable provided the links within the ecosystem are not forgotten.

8.2   RATE CONTROLLED GROWTH

8.2.1   Introduction

Rate controlled growth is described exactly by eqn (8.1.1), i.e.

$$dB/dt = r(t)B$$

where $r(t)$ is a continuous variable in time, i.e. there are no abrupt discontinuities in the value of $r(t)$. Restricting attention to the influence of physical factors only, the relation between such factors and growth can be considered as what may be called species-response curves. In examining the nature of such curves, McNaughton and Wolf (1973) distinguish between resources and regulators. In fresh waters, it is convenient to separate physical losses from other environmental regulators.

8.2.2   Resources

Resources determine the potential of an ecosystem. In the case of ecosystems where the base of the food web is photosynthesising plants, the resources are light energy and raw materials, i.e. the primary nutrients of inorganic carbon, nitrogen and phosphorus together with silica, calcium and trace elements. In the upper reaches of streams particularly, the resource that drives the system may be organic matter derived from the land.

The classic form of species-response curve for a resource is the Michaelis–Menten relationship between nutrient availability and growth rate (Fig. 8.4a). Although derived from totally different arguments, the response of species to light (Fig. 8.4b) has the same general form, i.e. there is a level of resource availability beyond which there is no increase in growth and, in the case of light, there may be inhibition if there is excess.

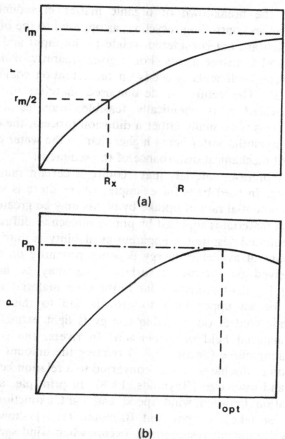

Fig. 8.4 Species response curves for resources. (a) Michaelis–Menten equation relating the growth rate, r, to resource availability, R, i.e., $r = r_m R/(R_* + R)$ where r_m is the growth rate when the resource is not limiting and R_* is the half-saturation constant. (b) General form of the relation between light intensity, I, and the rate of photosynthesis, P. The rate of photosynthesis may decline if the light intensity exceeds the optimum value, I_{opt}.

Referring back to Fig. 8.1, raw material availability is controlled by the amounts delivered from outside and recycled within the system. The amount delivered or the loading is determined by the discharge rate and complex relations between discharge and concentration which are not considered here. Raw materials may also be delivered by substances in rain falling directly on a water body. Most recycling results from the breakdown of organic matter in sediments. The physical processes controlling recycling, as opposed to the biochemical reactions which are not considered, relate to the input and output of material to and from sediments. For a given quantity of matter, the amount entering sediments depends on the retention coefficient (see Section 4.1.3). The return of decomposed material to the water column, provided it is chemically feasible, requires a transport mechanism. This is normally either a diffusion process, the concentration in the interstitial water being higher than in the water column, or some form of mechanical disturbance of the sediment.

Utilisation of raw materials may require a certain minimum turbulent mixing. In reed beds for example, where there is virtually no motion, the potential rate of uptake by plants may be greater than the rate at which material is supplied by purely molecular diffusion so that growth is inhibited despite the apparent availability of nutrients.

The amount of available energy depends primarily on the radiant energy received on the water surface. This may be modified by self-shading, i.e. light extinction due to the plant material itself so that there may be an upper limit to growth due to this alone. The hydrodynamic control on available energy is light extinction due to non-plant material held in suspension. In rivers, the conventional sediment rating curve (Section 5.2.3) relating the amount of material in suspension to discharge can be converted to a relation between light extinction and discharge (Reynolds, 1988). In principle, a somewhat similar relation between wind speed and light extinction could be developed for lakes. Carper and Bachman (1984) show that in a shallow lake, sediment resuspension occurs when wind speeds exceed critical values. This gives the lake equivalent of a sediment rating curve. Reynolds (1988), indeed, suggests that phytoplankton development in rivers, normally held to be limited by throughflow loss, may, in fact, be limited by available energy.

The resource type of species-response curve may also apply at higher trophic levels. The growth of feeding animals in response to available food supply frequently has a form similar to Fig. 8.4a and food availability may be controlled by physical conditions. Stream drift in rivers is an obvious example.

8.2.3   Regulators

Regulators are the form of species response for those physical factors which control the rate of growth and decay. The most obvious example is temperature which is not of direct concern here. The classic response to a regulating factor is the gradient model (Fig. 8.5) in which growth declines to either side of some optimum value although what happens at the extremes is not always clear. Response curves for temperature, for example, usually show a growth rate approaching zero asymptotically as the temperature falls but with a rapid plunge towards an upper lethal limit, the implication being extinction of the population if this lethal limit is exceeded.

The notion of regulators provides the starting point for examining the influence of other variables, particularly hydrodynamic effects. The classic model implies that growth rate is a maximum for constant environmental conditions. This may not be the case for aquatic environments where, over time, the growth rate may be a maximum for a sequence of different conditions. Consider, for example, the influence of hydraulic bed stress on the growth of bottom dwelling species. For a time, low stress with deposition and no resuspension is likely to give maximum growth—food supply is maintained and there is no erosion or other physical losses. After a period of time, however, the accumulation of deposited material may inhibit growth and there may be a longer term reduction of growth due to reduced material recycling. If this is correct, maximum growth over a period, therefore, would occur in a cycle of low stress, interspersed with periods of high

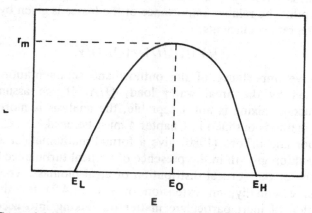

Fig. 8.5 General form of the relation between growth rate, r, and the value of an environmental variable, E. Jarvis (1976) suggests the following relation for the growth rate, r, in terms of the lower, E_L, optimum, E_o, and upper, E_H, limits of the variable, i.e., $r = a(E - E_L)(E_H - E)^b$ where $a = 1/(E_o - E_L)(E_H - E)^b$ and $b = (E_H - E_o)/(E_o - E_L)$.

stress. This example is hypothetical and may well be disproved by observation and experiment but it does relate to an important practical problem which is discussed in Chapter 9. If the flow in a river is regulated by a reservoir, what pattern of releases from the reservoir is optimal for the downstream river? The development of a time-varying species response analysis could well improve our understanding of the influence of physical conditions.

8.2.4 Physical Losses

Physical losses are a particular form of environmental regulator—the extent to which plankton population growth is controlled by losses from the system, specifically by outflow and sedimentation. If the assumption of instantaneous complete mixing is acceptable, the analysis is straightforward. In the case of outflow loss, the concentration decline following a sudden injection of tracer has been shown to be controlled by the theoretical retention time, T_r, i.e. $C = C_0 e^{-t/T_r}$ (eqn (4.1.6)). This is directly analogous to the equation for a falling population, i.e. $B = B_0 e^{-rt}$. The outflow loss rate is the reciprocal of the retention time and the usual presentation in terms of a species response curve is unnecessary.

An identical argument applies to loss by sedimentation if there is no resuspension. Equation (3.4.5) shows that the sedimentation loss rate coefficient is given by V_s/D where V_s is the settling velocity and D the water depth. The relative importance of the losses is given by the ratio of the loss rate coefficients, i.e.

$$(Q/V)/(V_s/D) = (Q/A)/V_s \qquad (8.2.1)$$

the relative importance of the outflow and sedimentation losses is determined by the areal water load, Q/A. If the assumption of instantaneous mixing is not acceptable, the analysis is more complex and the methods outlined in Chapter 4 must be used.

Schnoor and diToro (1980) give a formal mathematical analysis of phytoplankton growth in the presence of vertical turbulence leading to equations for the vertical distribution of algal biomass. Their starting point is, essentially, an extension of eqn (3.4.2) for the vertical distribution of inert particulate matter but taking into account algal growth rate and losses due to factors other than sinking. Separate solutions for the euphotic zone and a lower layer where growth is zero are given. Of particular interest is their equation for the Peclet

Number which is the ratio of the rates of advective and diffusive transport. This can be used to determine the combinations of growth, sinking and other loss rates in which phytoplankton will survive in different intensities of turbulence.

8.2.5 General Remarks

The notion of rate control is simple in principle, apart from the speculative comments about time-varying species response to hydro-dynamic variables. The growth rate varies directly with changes in the environmental variable, the effect of several variables being often assumed to be multiplicative, i.e. the net overall growth rate, r_*, is given by

$$r_* = r_1 \cdot r_2 \cdots r_n \qquad (8.2.2)$$

The problems are practical. Most species-response curves are non-linear so that, in order to determine the mean rate over a period, it is necessary to convert a time-varying physical factor to a series of rate values and then average the rates. Even more importantly, rates are, very often, not known, the most common observations being of the seasonal pattern of population change with concurrent observations of environmental variables. What is the relation between them, given that biological interactions are occurring at the same time? A population change with time when the growth rate is constant plots as a straight line if the ordinate has a logarithmic scale, increasing rates being concave upwards and decreasing rates concave downwards. Logarithmic plots, therefore, are useful in unravelling the various phases of growth. Once the timing of changes in growth rate have been identified, the influence of environmental variables and biological interactions is more easily detected.

8.3 EVENT CONTROLLED GROWTH

It was suggested earlier (Section 8.1.1) that instantaneous population falls can occur as a result of storms or other events. Despite its importance, the influence of events is not often considered in freshwater ecology and it is treated here in some detail. Later sections consider the response of species events and outline some possible applications. The concern here is with the influence of external events on what would, otherwise, be stable growth. It is not concerned with the implications of chaos theory, i.e. the possibility that the dynamics of the ecosystem are such that sudden changes in behaviour can occur without any external trigger mechanism.

8.3.1 The Nature of Events

The physical event (not the ecological consequences) may be discrete—the event occurs or not—or the exceedence of the threshold value of a continuous variable. Natural discrete events, such as landslides, are rare but they are more common in developed systems as with the sudden operation of a hydro-electric turbine. The exceedence of threshold values of wind or current speed, for example, is much more usual. Events are commonly random, i.e. their occurrence can only be expressed in terms of probability. The regular discharge of toxic material on Saturday nights, however, is not unknown to river managers. For continuous variables, the relation between magnitude and frequency of occurrence has to be taken into account. Extreme value analysis, one of the ways of defining this relation, has been illustrated earlier in connection with floods and droughts (Section 2.2.6).

The occurrence of a discrete event or of a value equal to or greater than the threshold value in the case of a continuous variable can be expressed directly in terms of its probability, p, or in terms of its recurrence interval or return period, T_i. The probability of a particular score in a single throw of a die is $1/6$. Alternatively, if the die is thrown often enough, the particular score occurs every sixth throw, on average, but it does not mean that the same score occurs every sixth throw. Formally, one can write $p = 1/T_i$. A throw with a die is, effectively, a time interval. The probability of an event occurring must have an associated time interval. With floods and droughts, for example, the usual time interval is 1 year, i.e. the flood probability defines the probability of it occurring in any particular year.

An important ecological question follows immediately. Given the probability or return period of an event, what is the probability that it will occur in a time interval T_*? By definition, the probability of the event occurring in the specified standard time interval is $1/T_i$ so that the probability of it not occurring is $1 - 1/T_i$. The probability of not occurring in two consecutive time intervals is the probability of not occurring in the first time interval times the probability of not occurring in the second time interval. The probability of not occurring in T_* time intervals, therefore, is given by

$$p(\text{not } T_*) = (1 - 1/T_i)^{T_*} \qquad (8.3.1)$$

Hence the probability of the event occurring in T_* intervals, $p(T_*)$, is

$$p(T_*) = 1 - (1 - 1/T_i)^{T_*} \qquad (8.3.2)$$

Equation (8.3.2) is often presented in an alternative form in what is termed, rather loosely, as the Poisson approximation. This follows from the fact that, provided T_i is large compared to T_*, it is possible to write

$$(1 - 1/T_i)^{T_*} = [(1 - 1/T_i)^{T_i}]^{T_*/T_i} \rightarrow e^{-T_*/T_i},$$

so that

$$p(T_*) = 1 - e^{-T_*/T_i} \qquad (8.3.3)$$

If $T_* = T_i$, $p(T_*) = 0.632$, i.e. there is a 63 per cent chance of an event occurring within a period of time equal to the return period. Also, there is a 1 per cent chance that the event will not have occurred in a period of time equal to 4.6 times the recurrence interval.

So far, eqn (8.3.2) has been used to determine the probability that an event occurs. The equation, in fact, defines one minus the probability of no events occurring so that the possibility of more than one event has not been excluded although the Poisson approximation assumes that the probability of more than one event is so small that it can be neglected. Equations for the probability of 2, 3 or more events in a specified time interval are given by Smith (1987).

8.3.2 Species Response to Events

A population's response to events has to be interpreted in terms of two parameters—the interval between events and the magnitude of the population fall due to the event (Fig. 8.6). The simplest possible assumptions about the nature of the population fall are that either the population is reduced to some constant value, irrespective of the pre-event population, or that the post-event population is a fraction of the original value, i.e. the number of survivors is proportional to the pre-event population.

It is possible to envisage the influence of events as limiting cases of rate controlled growth. Referring back to the species response curve for a regulating variable (Fig. 8.5), it is possible to envisage extending

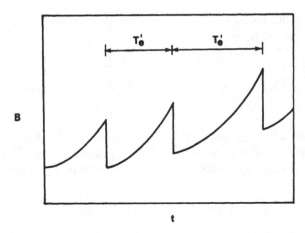

Fig. 8.6 Biomass change due to external events. The controlling factors are the interval between events, T_e', the magnitude of the biomass decline as a result of the event and the growth rate between events.

beyond the upper, lethal limit and plunging rapidly towards infinite negative growth rates, causing instantaneous death of some proportion of the population. Frequent, small events have the same effect as rate control. If the events are frequent enough, the influence of random events will, effectively, be the same as that of events occurring at regular intervals. Small variations in timing will not be significant and, with constant intervals, the influence of events results in an exponential decline expressed as a difference equation (Fig. 8.7). Consider a population that would be constant in the absence of events, where the interval between events is T_e' and where the fractional loss due to an event is α. If r_e is the effective loss rate due to events, then, in terms of rate control,

$$B = B_o e^{-r_e T_e'} \qquad (8.3.4)$$

But, in one time step, $B = \alpha B_o$, so that

$$r_e = \ln \alpha / T_e' \qquad (8.3.5)$$

If r is the growth rate in the absence of events, then the combined effect of growth and losses due to frequent small events is given by

$$B = B_o e^{(r - r_e)t} \qquad (8.3.6)$$

In order to explore the influence of events more fully, Smith (1987) examines how a species growing according to a logistic equation

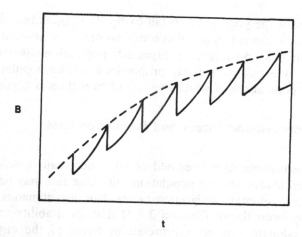

Fig. 8.7 Biomass change due to frequent, small events interpreted as rate control. As the interval between events and the loss due to the event become smaller, event control is, effectively, reduced to rate control. The equivalent net growth rate, indicated by the dashed line, is given by eqns (8.3.5) and (8.3.6).

responds to events, the logistic equation being taken as the simplest possible model of growth. Although random event intervals are much more likely in reality, examination of what happens with constant intervals does shed some light on the influence of events. The main feature of growth with constant event intervals is that, irrespective of the initial conditions, the growth pattern either settles down to a steady cycle, i.e. the post-event population is constant or the species becomes extinct. For steady cycles, the ratio of the mean population over a cycle to the carrying capacity gives a measure of the population suppression due to events.

What happens with random intervals between is more difficult to analyse but, given the logistic and event parameters, population-risk curves can be constructed. These give the relation between the population at any time and the probability that this population could be rendered extinct by events. It is assumed there is a critical, minimum population, B_c, at which the population becomes extinct for other biological reasons. If the population is less than B_c/α where α is the fractional population loss due to an event, then it can be extinguished by one or more events in the time it takes to grow to B_c/α. The probability of an event occurring in this time can be calculated using eqn (8.3.2). At higher populations, several events must occur for the population to fall to B_c. The population-risk curve is built up by summing all possible combinations of events and times of exposure to risk. As might be expected, population-risk curves are hyperbolic—the risk is inversely proportional to the population—but there exist populations at which the risk of extinction is negligible.

8.3.3 Hydrodynamic Stress and Population Loss

Physical disturbance of a river bed or lake shore can cause a sudden loss of benthic invertebrate populations although this may be, at least initially, an involuntary emigration rather than instantaneous death. It has already been shown (Section 3.4.4) that the stability and movement of sediment can be interpreted in terms of the entrainment function, θ. The possibility exists, therefore, of a relationship between the population loss, α, and the entrainment function. This is simply a hypothesis unsupported, as far as is known, by any experimental evidence.

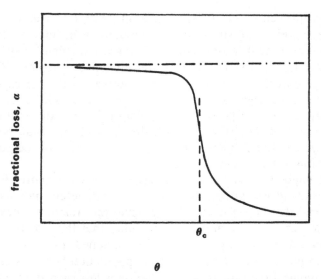

Fig. 8.8 Population loss caused by the disturbance of sediment. Figure shows the schematic relation between the entrainment function, θ, and the fractional population loss, α. The bulk of the loss occurs at a critical value of θ when the disturbance of the sediment is so great that evasive action by invertebrates becomes increasingly difficult.

Consideration is limited to coarse bed material where the value of θ required to carry material into suspension is appreciably greater than that required to initiate bed movement. The hypothesis is based on the assumption that three phases occur in the θ–α relation (Fig. 8.8), viz.

—low values of θ which cause small population loss and which can be treated as rate control;

—sudden large population loss at a critical value of θ where the bed is so disturbed that the habitat is partly destroyed and where evasive action by invertebrates becomes increasingly difficult;

—relatively slow decrease in the remaining population when θ is greater than this critical value and which, again, could be treated as rate control.

If the prime cause of population loss is partial destruction of the habitat, then the critical value of θ will be determined by the nature of the sediment. If this is the case, then it is reasonable to assume that the critical value of θ will be greater than that for the onset of bed movement (0·45) but somewhat less than Bagnold's criterion for fully developed suspension (2·50). Using the methods outlined in Section 5.3.2, the critical value of the entrainment function can be associated with a flow rate and, thus, with a return period.

8.3.4 Population Loss Due to the Sudden Discharge of Toxin

This second example shows how the response of an organism to the sudden discharge of a toxic substance into a water body can be interpreted in terms of rate control. The physical conditions in the water body are the simplest possible—instantaneous mixing in a water body of constant volume and with constant throughflow. Two stages, acute and chronic, in the organism's response are assumed (see, for example, Holdgate, 1979). The effects of the acute stage are irreversible and usually fatal. The simplest possible assumption is that there is a linear relation between toxin concentration and death rate. Also, there is a minimum response or contact time before any deaths occur. During the chronic stage, no further deaths occur but growth is inhibited by the toxin. This sublethal effect can be interpreted as a negative resource, i.e. by an inverted form of Michaelis–Menten kinetics—the higher the concentration, the lower the growth rate.

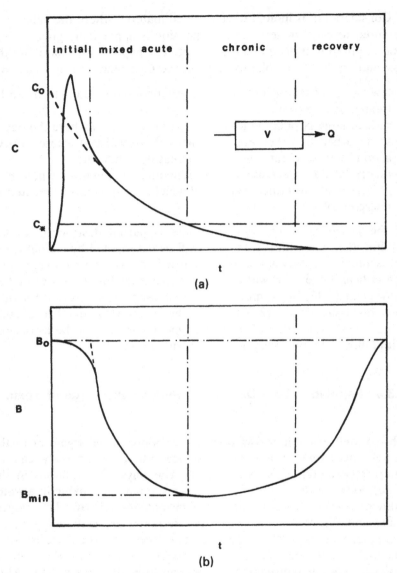

(a)

(b)

Fig. 8.9 Schematic outline of an organism's response to the sudden injection of a toxic substance into a water body. (a) Concentration changes in the water body showing the different phases of the organism's response. C_* is the limit of the acute toxicity. (b) Corresponding population changes. Some deaths are likely in poorly mixed pockets during the initial phase.

The overall organism response can be seen as four phases (Fig. 8.9), viz.

—Initial phase which covers the minimum exposure time before lethal effects can be detected and the time required for the toxic material to become fully mixed, the two times being assumed equal.

—Mixed acute phase during which deaths occur until the toxic concentration falls below the lethal limit as a result of dilution.

—Chronic phase during which growth is suppressed until the toxic concentration becomes insignificant.

—Recovery phase, i.e. the time required for the population to grow to its original, pre-injection level.

Preliminary analysis indicates that the duration of the various phases (other than the initial) can be expressed as multiples of the theoretical retention time, V/Q.

The population loss, α, is determined by what happens during the mixed acute phase, i.e. $\alpha = B_{min}/B_o$. The value of α is a complex function of the hydrological factors controlling the rate of dilution and of the toxicity parameters for the organism concerned. Clearly, the population fall is not instantaneous but the duration of the mixed acute phase is normally short compared to the combined length of the chronic and recovery phases. The recurrence interval is dependent on external factors.

8.4 SPACE CONTROLLED GROWTH

8.4.1 Spatial Variation in Plankton Growth

Spatial variation in a water body implies more than the direct consequences of the hydraulic structure within it. Chemical composition and plankton density, for example, may vary for other reasons. In rivers and on lake beds, spatial differentiation has been described already and the concern here is with planktonic organisms whose distribution is less restricted by solid boundaries. How adequate is the 'representative vertical column' of the idealised lake ecosystem?

George and Heaney (1978) distinguish two primary causes of horizontal variation in plankton distribution, viz.

—factors causing local changes in growth rate, such as nutrient availability and temperature differences,
—spatial redistribution of the population as a result of water movement.

Local changes in growth rate occur when the equalisation or mixing time is greater than the time scale implied by the growth rate. Growth rates close to inflows and in sheltered bays, for example, may be higher than those in open water but such differences usually disappear with increased mixing at higher wind speeds. Differences in the time scales for growth and mixing can result in long lasting, large scale horizontal variations in the open sea (Platt and Denman, 1975). The distances involved and the presence of boundaries, however, means that this is an unlikely cause in all but the largest lakes. More permanent spatial variation in growth rate in small to medium sized lakes is more likely to be due to the relative positions of the major inflows and the outflow. If short circuiting occurs so that a system with backflow is the appropriate compartment model structure (see Section 4.3.4), then nutrient concentration differences may be sustained despite wind induced mixing.

The redistribution of plankton directly reflects the hydraulic structure and scales of turbulence within a water body. Horizontal and vertical variations, usually related, occur and both phyto- and zooplankton are affected. Both forms of plankton are passive in relation to horizontal movement but vertical movement is subject to some control. The vertical distribution of phytoplankton is controlled by buoyancy and vertical turbulence while vertical zooplankton movement is primarily a response to light. Spatial variation on a basin wide scale where the two dimensional circulation is that of Fig. 6.13 is well illustrated by George and Edwards (1976). Positively buoyant organisms accumulate in the downwelling and negatively buoyant ones in the upwelling, neutrally buoyant organisms being uniformly distributed throughout. Positively buoyant algae can also accumulate near the surface in light winds but becomes more uniform vertically when the wind speed exceeds about $4 \, m \, s^{-1}$. This is consistent with the observed increase in the downward transport of kinetic energy at approximately the same wind speed. The surface concentration downwind can be very high and, if a gradient towards the outflow is maintained for any length of time, the loss of plankton down river may be much greater than if uniform distribution is assumed.

8.4.2 Habitat Availability

The notion of carrying capacity is well established in ecological theory—'the upper limit of a population in an environment, beyond which no major increase can occur. Represented by the upper asymptote of the sigmoid curve of population growth' (Chambers, 1974). The difficulty arises when attempting to give carrying capacity a physically real interpretation.

Carrying capacity can be viewed as the limit of rate controlled growth. Phytoplankton in a eutrophic lake is an obvious example. Once a certain algal density is reached, further growth so increases the light extinction that the population becomes self-limiting. There is no direct spatial limit. Event control can also impose an upper limit. Alternatively, there may be an upper limit to a population because of the lack of suitable habitat even although the rate and event control conditions are far from limiting. This is particularly the case with rooted macrophytes and bottom dwelling species on river beds and lake shores.

Formally, in the case of growth restricted by habitat availability, the fundamental growth equation can be modified to include a habitat coefficient, χ, i.e.

$$\mathrm{d}B/\mathrm{d}t = \chi r B \qquad (8.4.1)$$

If $\chi = B - B_*/B_*$, i.e. where it is determined by the difference between the current population and the carrying capacity, the conventional logistic equation is obtained. Such a relation is the simplest possible. Not only is there the question of determining the upper limit, B_*, but also there is a variety of ways in which habitat availability may inhibit growth at lower populations (Fig. 8.10). Equation (8.4.1) suggests the possibility of a number of growth models incorporating both population restriction due to habitat availability and time dependent growth rates.

The real problem is the estimation of carrying capacity. Can an upper population limit be obtained from knowledge of site characteristics? Naively, an upper limit can be measured as the available area of suitable habitat divided by the area required per individual although the area required per individual may itself vary with rate controlled

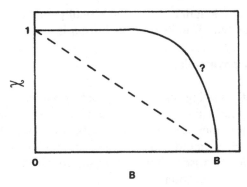

Fig. 8.10 The limitation of population growth by habitat availability. The figure shows possible relations between the habitat coefficient, χ, and the population, B. Dashed line is equivalent to the logistic equation, i.e., $\chi = 1 - B/B_*$, so that $dB/dt = rB(1 - B/B_*)$ where B_* is the carrying capacity. Other forms of restriction due to habitat availability are possible.

factors. Despite the difficulties involved, the ability to distinguish an upper limit due to habitat availability from the limits imposed by rate and event control is essential for a proper understanding of the influence of physical conditions.

8.5 FINAL REMARKS

The logic behind assessing the influence of physical conditions is simple. For a species to be successful ecologically at a particular location, it must be able to:

—reach that location,
—grow successfully, i.e. the gains must be greater than the losses,
—reproduce successfully and establish the next generation.

Mobility and distribution in the case of migratory fish can be understood from a straightforward examination of river structure—are there natural or artificial barriers to movement? Many other questions relating to species mobility cannot be explained by present day physical structure. Fish occur in river reaches where migration by water appears impossible and, while some species have aerial phases in their life cycle, present distribution is often difficult to explain. Birds often play an important role in speculations about species distribution.

The main concern here, however, are the conditions for successful growth and reproduction. The physical requirements for success are determined by habitat availability, susceptibility to events and the existence of a positive net growth rate although concentrating on these may not provide an immediate, direct route to assessing the influence of physical conditions. In the discussion of rivers, for example, considerable emphasis is placed on the importance of hydrodynamic conditions but a species response curve, directly relating growth rate to the state of flow, may not be feasible. Too many different processes may be involved for such a relation to exist.

Concentrating on the fundamental controls on growth rate does provide a logical basis for formulating hypotheses and interpreting the results of observations. In addition, it provides the means of extrapolating results to other sites and environments. An observed change in species density following flow regulation, for example, cannot be taken as a general rule if the processes causing that change are not understood. A similar change in a different environment could have different results.

It is possible to envisage a number of measures of the properties of ecosystems relating to the influence of physical conditions. One example is environmental susceptibility—the extent to which an ecosystem is affected by environmental factors. In this sense, a deep lake is likely to be less susceptible than a small stream whose environmental characteristics closely follow current weather conditions. Susceptibility can be imagined as a non-biological facet of Holling's (1973) concept of resilience—the ability of an ecosystem to withstand stress of any kind without permanent damage to its structure and dynamics. Nevertheless, if any of these properties are to be more than abstract concepts, the only route to any meaningful measure is through rate, event and space control.

9

Synthesis

9.1 BACKGROUND

So far, attention has been focussed on the role of all forms of water movement on the ecology of fresh waters. Some features of aquatic habitats and the way these influence the growth of plants and animals have been examined in isolation. In addition, because few river basins are in a pristine state, the changes resulting from man's activity have been reviewed. What has to be considered now is how to link these separate elements together, not only to improve our understanding of freshwater ecology but also as a contribution to the management of river basins.

9.1.1 Interests Involved in River Basins

The process of integrating the different strands involved is partly a technical question and this is discussed in detail below. But it is also a conceptual issue. Even the technical questions cannot be separated from the different views held about the function of river basins and the different traditions and emphases of the various sciences involved. Apart from deserts and polar regions, almost the entire land surface of the earth is within the catchment of a river basin. The potential influences on a river basin are, therefore, virtually infinite. For simplicity, these are lumped as land use interests—agriculture, forestry and the construction industry, for example—which treat a river basin as a transport or waste disposal system.

The variety of interests impinging on river basins are indicated on Fig. 9.1 and it is wise to separate vested interests from public interest and concern. The dominant vested interests are those which view a

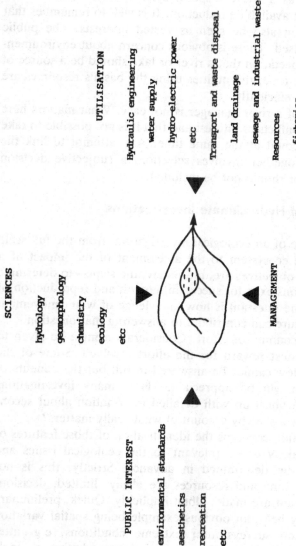

UTILISATION

Hydraulic engineering
water supply
hydro-electric power
etc

Transport and waste disposal
land drainage
sewage and industrial waste

Resources
fisheries
etc

SCIENCES

hydrology
geomorphology
chemistry
ecology
etc

MANAGEMENT

resolving conflicts
optimisation
legal and financial
technical

PUBLIC INTEREST

environmental standards
aesthetics
recreation
etc

Fig. 9.1 Interests involved in river basins.

river basin as a resource to be exploited and developed. Obvious examples are: all those activities associated with hydraulic engineering such as water supply and hydro-electric power generation; activities which view a river as a waste disposal system; fisheries and other activities that harvest available production. It is well to remember that scientific interests can also be seen as vested interests. The public interest can be confused. There is obvious concern about environmental issues and the expectation that a river or lake should be a source of delight. At the same time, the utilities using the basin's resources are expected to function efficiently.

Clearly, the task of a river manager is not easy. What matters here is to recognise the conflicts of interest and that it is not possible to take all interests into account at the same time. Any attempt to link the separate elements together involves selection, a subjective decision about what should or should not be included.

9.1.2 The Logic of Hydroclimate Investigations

Whatever the nature of an ecological investigation, from the full scale study of an aquatic ecosystem to the assessment of the impact of a specific change, the objectives are, ultimately, the same—to determine whether the conditions exist for successful growth and reproduction of different species. The question is how knowledge of water movement and hydraulic structure can contribute to answering that question.

An essential is economy of effort. Consideration must be given to what gives the greatest reward for the effort involved. Some of the questions posed below cannot be answered in full but the benefits of having asked them can be appreciable. How many investigations (including my own) finish up with detailed information about secondary issues and only a sketchy account of what really matters?

An obvious question concerns the identification of those features of the hydroclimate likely to be relevant to the ecological issues and whether they can be determined in advance. Strictly, this is not possible but, since time and resources are always limited, decisions about what is relevant are made, albeit implicitly. Quick, preliminary observations clearly help, an obvious example being spatial variation in a lake. Plankton surveys after extreme conditions, e.g. after prolonged calm or windy weather, may indicate whether a single representative column is adequate. Flexibility is essential—a willingness to modify programmes in the light of experience gained.

Another issue is the degree of detail and the level of accuracy required. In any investigation, the error and uncertainty in all its components should be the same. What level of physical investigation is appropriate to the accuracy and detail of the ecological work? Sensitivity analysis, without any mathematical formality, may identify factors that have a potentially dominant influence on the topics of interest. Simulating normal conditions may well be less important than assessing the risk of some threshold value being exceeded or of some event occurring.

Most other questions expand the theme of the degree of detail and accuracy required. The appropriate time and space scales are obvious examples. Assuming that the biological sampling frequency is already decided, the choice of time and length scales for environmental measurements—the required degree of averaging—is not easy and is linked to difficulties caused by non-linearities. A simple, run-of-wind anemometer, read at the time of a sampling visit, gives an accurate indication of mean wind speed since the previous visit but gives a poor indication of wind stress. It is important to distinguish between indicators of average hydroclimate and physical factors directly related to the topics being studied.

Linked to the question of time scales is the use of data already available. Continuous, accurate streamflow measuring stations, for example, are operated in many rivers. Establishing a correlation between flow at the standard station and an experimental site is relatively simple. If what is required, for example, is the average flow over 5 day intervals, then such correlations are useful but the accuracy declines as the time interval is reduced.

Spatial limits and scales must also be defined. In systems theory terms, it is important to be clear about what constitutes the system under investigation and what, in these terms, is environment—inputs and outputs that are not of direct concern. Spatial scales within the system are, obviously, related to time variability. There is no point in looking for spatial variation if the variability at a point is greater.

Instrument availability, although not discussed here, is clearly important. Associated with this is the question of whether some factor should be measured directly or whether a predicted value is adequate although this does not arise if the water body under investigation does not yet exist as can happen with reservoir proposals. Clearly, a balance has to be found between instrument availability, costs, available time and the confidence that can be placed in predicted values. The discussion of models below suggests a number of instances where a choice between observed and predicted values is possible. It is important to ensure physical investigations are formulated as experiments and not simply as data collection. Calculated values can be used to form an hypothesis which can be tested by limited, critical observations, successful confirmation increasing confidence in the predicted values.

9.1.3 Modelling Aquatic Systems

One way of linking different components together is the construction of mathematical models—a representation, in mathematical terms, of some of the processes involved in a particular system. It is worth commenting on the use of the word model in this context. It is often assumed to mean a very large investigation, mathematically obscure to the non-specialist and where the results of interest cannot be found in the pile of computer print-outs. It need not be so. A description of the cross-sectional form, the bed particle size distribution and the form of the stage-discharge curve provide the basis of a useful model river behaviour at a point. Models can be used for management where the object is to reproduce some of the characteristics of a real life system and, in particular, to predict the consequences of changing the variables that control system behaviour. They can also be used as research tools, i.e. as experiments or numerical trials in which the implications of different assumptions and hypotheses can be explored.

It is not the intention here to develop specific models that are more or less ready to run but rather to show how the material of previous chapters can be used to form the components of model systems, appropriate to the objectives of a particular investigation. The synthesis of model rivers and lakes is considered before outlining possible ways of applying rate, event and space control in such model systems. The theme of the Preface is reiterated, viz. that the object is to stimulate ecologists to consider the implications of environmental conditions rather than to suggest that these are understood already.

9.2 RIVER MODELS

Longitudinal variation and dependence on upstream processes are the dominant features of rivers. A river model, therefore, must indicate how river characteristics vary along its length. Account must also be taken of time variations, usually expressed as variation with discharge. Various statistical descriptions of river ecology relate species distribution to physical features such as slope and channel width. A requirement of any model is that it should make it possible to use such relations in order to provide the means of linking species distribution to catchment topography and climate. The main objectives, nevertheless, are to consider ways of constructing models that express the influence of water movement and then to consider how rate, event and space control can be applied in such models.

9.2.1 River Description

Rivers are indeterminate in the hydraulic sense, i.e. they are free to adjust more characteristics, slope, width, depth and particle size, for example, than the number of equations that can be developed on purely physical arguments. What can be exploited is the fact that the river channel is formed by flowing water so that there are well established empirical relationships between flow rate and channel dimensions (see Section 5.1.3). Dependence on empirical relationships presents a number of problems. The accuracy and validity of the relationships themselves must always be borne in mind and their meaning has to be considered. Empirical equations define average relationships. For example, in the relation between discharge and channel width, the estimated width is the average for a length of river having constant flow. Any modelling of longitudinal variation that uses empirical relations is based on the assumption that changes over different lengths are greater than the variation within a river length.

The essential structure of a descriptive river model for a basin containing no lakes of any size is shown in Fig. 9.2. This simply shows the way in which material from earlier chapters can be linked together. The two initial boxes are based on Chapter 2 and specify the nature of the river, i.e. how the independent or driving variables related to catchment characteristics, climate and streamflow vary along the length of a river. The other boxes are based on Chapter 4 and explore the implications for river structure and dynamics of the specified river features.

Some examples of the type of output obtainable from such models are shown in Fig. 9.3. These are based entirely on computed values,

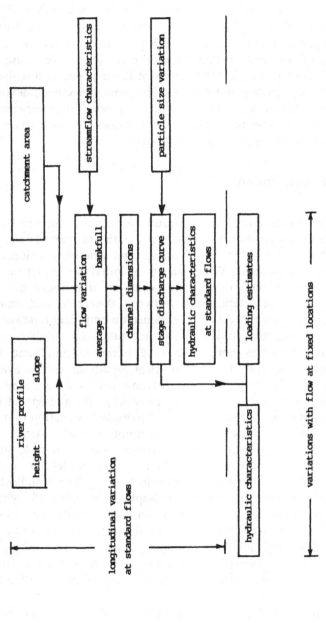

Fig. 9.2 Structure of a descriptive river model showing the way material from earlier chapters can be linked together to give an overall description of a river system.

i.e. the entire river is described by a few parameters relating to catchment form, the variation of rainfall with altitude and by the condition that the cross-sectional form of the channel is parabolic throughout. The entire illustrated output could be measured directly, the only difficulty being that this would be expensive and time consuming if the entire river is being investigated. A limited programme of direct measurement obviously increases confidence in predicted values, for example, by comparing observed and predicted cross-sections at various points. Rivers are easily seen and their dimensions perceived without the aid of instruments. There is a tendency to expect greater accuracy in what we can see directly compared to what is deduced from instrumental observation. In addition, total reliance on measured values implies, effectively, that every river is unique and that no generalisations are possible.

The results illustrated in Fig. 9.3 are very limited. For example information on sediment sizes on the river bed could be incorporated. Here again, a combination of direct observation and analysis may be effective—is there, for example, systematic variation in the coefficient of fixation, d/S? Given sediment sizes, a river can be seen as a sequence of habitats whose characteristics are determined by the nature of the boundary layer flow. It is possible to summarise the features of this flow in terms of Fig. 5.14. This is a plot of bed slope against the ratio of water depth to bed particle size and was originally derived in order to characterise river stability. The nature of the water surface corresponding to any position on the graph is characteristic and an approximate scale, analogous to the Beaufort Scale for the state of the seas, can be derived.

This view of a river as a sequence of habitats with different hydrodynamic characteristics seems preferable to other methods of subdivision that have been proposed. There is no implied relation between stream order number and environmental conditions, the division into eroding, transporting and depositing regimes is not totally satisfactory and divisions based on hypothetical species distribution are arbitrary and subjective.

9.2.2 Transport and Storage

So far, only the hydrodynamic features of river habitats have been considered. Equally important is the downstream transport of material and how it is modified by reactions within the river. Sophisticated

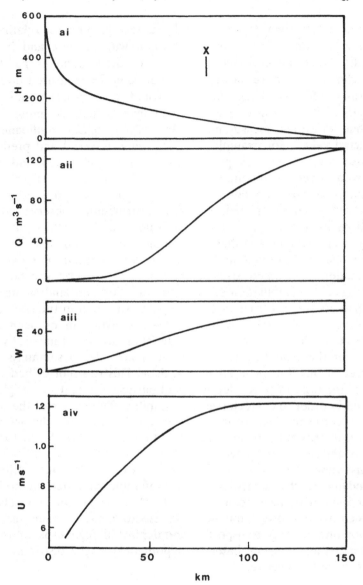

Fig. 9.3 Sample output from a descriptive river model. (a) Variation in river features with distance from the source. ai longitudinal profile; aii average flow; aiii channel width at average flow; aiv mean velocity in cross section at average flow. (b) Cross sectional form at point X, showing the water level corresponding to average flow. (c) Variations with discharge at point X. ci water depth; cii mean velocity in cross section.

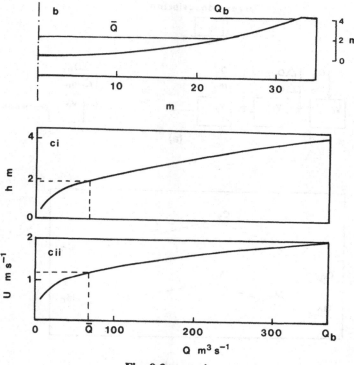

Fig. 9.3—*contd.*

techniques are available for predicting downstream water quality changes below outfalls (see, for example, Rinaldi *et al.*, 1979). The object here is much simpler. Given the pattern of inputs and sources, what is the longitudinal distribution of material in the water and retained on the river bed and what is delivered downstream, ultimately to the sea?

It is possible to envisage a river as a cascade—a linked series of basins but where the downstream increase in discharge and the sources of material appear as additional inputs between basins (Fig. 9.4a). There need not be additional inputs at every link. A steady state, mass balance equation, including internal reactions, can be solved for each basin in turn, the input being the output from the one above plus what is added. The principle is simple and unoriginal, the problems, which are related, being to decide the volume of each cell and the nature of the mixing within each cell. The channel cross-sections are known

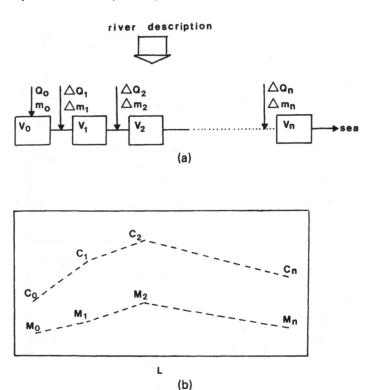

Fig. 9.4   Schematic model of transport and storage in a river. (a) Simplified view of a river as a cascade of basins in series. (b) Typical output showing changes in river water concentration, C, and the amounts retained in each basin, M.

from the descriptive model so that determining the volume is really a matter of deciding the length. The strong directional flow in rivers means that mixing is more likely to be nearer to plug flow than instantaneous mixing but intermediate reaction efficiencies for the given conditions can be used (see Section 4.2.3). The selection of cell length depends on a number of factors: the river configuration; the degree of detail required; the throughflow rate constant, Q/V, compared to the internal reaction rate; the nature of the mixing process assumed. It is wise to run several versions of any model with different cell volumes and mixing characteristics to check the influence of model structure on the results.

The results obtained for flows with a particular frequency of occurrence are the longitudinal change in concentration in the water and in the amount accumulated in the river (Fig. 9.4*b*). The mean cell concentration for plug flow can be taken as the mean of the inflow and outflow concentrations. By running the model with flows exceeded different percentages of the time, the effects of dilution and of any links between input amounts and flow can be tested and the probability of critical concentrations being exceeded can be estimated. The potential for increasing the complexity in both the processes described and the mathematics is obvious.

9.2.3 River Ecology

In order to apply the principles of rate, event and space control to rivers, it is necessary to have some picture or model of the organisation of a river ecosystem. Given the linear structure of a river, the causes of the longitudinal variation in species distribution are an obvious starting point. The assumptions are that species will only be present when there is positive net growth, i.e. gains over time are greater than losses, that their density is proportional to the net growth and that physical conditions determine the potential species distribution. Biological interactions, such as competition, are only likely to cause exclusion of a species.

The application of rate control is confounded by the fact that environmental variables vary with discharge as well as longitudinally. Before longitudinal variation in growth can be considered, a representative rate for a particular site must be defined. This raises problems as most species response curves are non-linear—how to obtain the best estimate of the representation rate at a site? This may be the rate that occurs most of the time or, and this may not be the same, the rate at which most growth occurs, i.e. the rate at which the product of rate and duration of growth at that rate is a maximum.

Assuming that representative site values can be established, growth in terms of river distance can be examined. Figure 9.5 shows the combined effect of a single resource limitation and a single environmental regulator. The resource control follows Michaelis–Menten kinetics (Fig. 8.4) and the influence of the regulator is based on Jarvis's equation (Fig. 8.5). For simplicity, both environmental variables are assumed to increase linearly from source to mouth. The hypothetical species distribution is based on the assumption that the relative growth rate must be greater than 0·5 for the gains to exceed

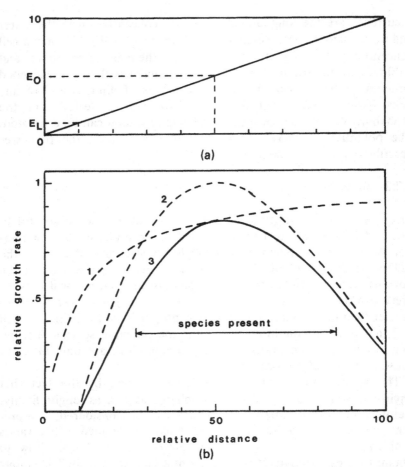

Fig. 9.5 Schematic illustration of the influence of rate control on species distribution in a river. Growth rate is assumed to be controlled by a single resource and a single environmental regulator. (a) The longitudinal variation in both controlling variables. E_L is the lower limit for the environmental variable and E_o the optimum value. (b) Longitudinal variation in growth rates. Line 1 indicates the resource controlled growth, line 2 the regulator and line 3 the combined effect. The presence of the species is based on the assumption that the growth rate must be greater than 0·5 for the gains to exceed the losses.

the losses. The example is purely illustrative but it does emphasise that sharp divisions in species distribution are unlikely.

Event control—sudden losses of population due to specific events— can also affect species distribution. A likely cause is habitat distur- bance which reflects both the hydrological regime and, ultimately, geology in how it influences bed particle size, but sudden losses can also be caused by the occurrence of lethal temperatures or the accidental discharge of pollutants. It is a requirement of the descriptive modelling to establish the probability of such events at points along a river.

In regions where the event probability and resulting losses are low, a quasi-steady state ecology, determined primarily by rate control, can be established. Where the event probability is high, no steady state exists, populations fluctuate erratically and, if normal conditions can be envisaged at all, it is the recovery phase following an earlier event. Population densities are suppressed by the repeated, event related, losses.

Event frequency has important biological implications which are only mentioned here. Animal species are often separated into the so-called K- and r-strategists (Southwood, 1981) and the relevance here is obvious. K-strategists are slow growing species that exploit stable habitats, they are larger, longer lived animals and their behaviour emphasises defence mechanisms and competitive ability. The opposite behaviour is demonstrated by the colonising r-strategists with a rapid growth and utilising whatever resources are available. Changing species composition during the recovery phase between events appears possible. Effectively, there is biological succession within a river and, if the undisturbed period is long enough, a stable, climax species composition may emerge.

Spatial variation is implicit in the structure of a river model so that the only aspect of space control that needs to be considered is habitat availability—the extent to which this limits growth and distribution. As with rate control, the available habitat varies with discharge and may be obtainable from an examination of the variation of channel characteristics with flow rate. Obtaining an appropriate value for a site, however, is not just a statistical problem. The capacity of a species to make use of available habitat, i.e. for increased habitat space to result in increased population, is similar to May's (1981) analysis of logistic growth with a variable carrying capacity. If the population response time, as measured by the reciprocal of the growth rate, is long compared to the environmental periodicity, then the population is stable and low. Conversely, if the response time is short, the population tracks the environmental variations. Figure 2.19 suggests that this distinction is partly a matter of river size—fluctuations in flow may be less in very large rivers—but seasonal variations are also important. If the fluctuations are so severe that the habitat is destroyed, then space control is, effectively, replaced by event control.

The stated objective—the causes of longitudinal variation in species distribution—results from all three controls acting, sometimes simultaneously and sometimes sequentially. This outline is dominated by its emphasis on physical factors to the neglect of biological interactions. All that it is intended to do is to provide a means whereby what needs to be known about the physical structure and dynamics of a river can be assessed.

9.3 LAKE MODELS

9.3.1 Introduction

It is clear that the water movement and associated hydraulic structure in a lake is complex and not fully understood. In addition, few ecological investigations cover the whole range of processes from the photosynthesis of plant material to the feeding behaviour of the top predators. A decision must be made as to what features are to be investigated. In contrast to rivers, therefore, there can be no single model that predicts lake hydroclimate nor any universal definition of objectives in ecological investigations. No attempt is made, therefore, to define specific models. Lake hydroclimate investigations can only be based on searching for what gives the greatest reward for effort. The logic is outlined in Fig. 9.6.

9.3.2 Basic Features of Lake Hydroclimate

The contents of Chapter 6 indicate that an approximate assessment of the general features of water movement can be made given a bathymetric map and knowledge of the wind and flow regimes to which the lake is subject. Obvious examples are the dynamic ratio, $\sqrt{A_L}/\bar{D}$, which determines the importance of wave action and the extent of disturbed lake bed, and the seasonal variation in the energy inputs from wind and inflow which indicate the relative importance of wind and flow driven circulation. In addition, a tentative sketch of the hydraulic structure under different wind conditions should be possible. It is assumed that the thermal structure is known since temperature

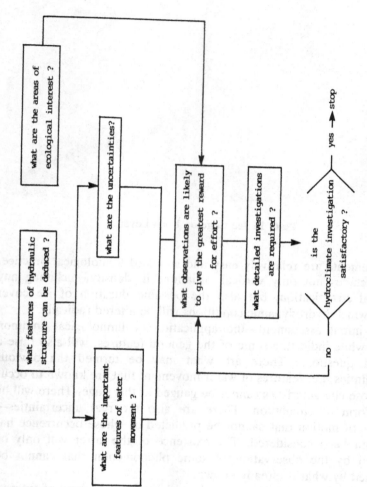

Fig. 9.6 The logic of lake hydroclimate investigations.

Plate 15 Ice ridge on Loch Leven.

measurements are relatively easy and standard limnological practice. This includes not only vertical variations in density, including any chemical stratification, but also the possible duration of ice cover during which hydrodynamic conditions will be altered radically.

This initial assessment—the application of limnological common sense—while indicating some of the general features, will also raise a host of questions. There are what may be termed the obvious uncertainties, i.e. features of water movement that are known to occur but whose characteristics cannot be gauged at this stage. There will be some form of circulation. There are also the real uncertainties— features of motion that cannot be predicted or whose occurrence has not even been considered. The existence of the latter will only be detected by the observation of some phenomenon that cannot be explained by what is already known.

Because of the diversity of ecological topics, the areas of interest should be identified before proceeding further since these largely determine the hydroclimate investigations required. As in Chapter 8, the distinction can be made between the view of a lake as a chemical reactor and environmental influence on lake habitats (see below). The rest of the procedure is iterative, finding the balance between on-lake observations and predictions that can be made in the comfort of the laboratory. Chapter 6 also indicates a number of features of lake hydrodynamics where specific predictive procedures, models, can be developed. Many features of wave action and seiche motion, for example, can be predicted with reasonable confidence.

9.3.3 Lake Ecology

The simple but long standing view of a lake as a chemical reactor has been in terms of what are often referred to as loading models. Starting with Vollenweider (1968) but developed further by Vollenweider (1975) and Dillon and Rigler (1975), various relations between nutrient input, throughflow and phytoplankton density have been established. Their merit is their practical value. For a given lake and catchment, the consequences of any nutrient input can be expressed as changes in lake water quality, i.e. as plankton density or trophic status. Vollenweider's initial model took no account of throughflow but even the later versions are based on a very simple view of plankton ecology. Steady state conditions are assumed, only physical losses through outflow and sedimentation are considered and an empirical relation is used to establish the link between available phosphorus and plankton density.

Other features of the hydroclimate clearly have an influence on open water habitats and the relation between nutrient input and algal growth. The compartment model structure of the lake and the resulting residence time distribution may have an influence on reaction efficiency. In shallow lakes with extensive disturbed areas of bed, appreciable quantities of nutrients may be recycled so that there is no direct link between input and availability. A third question concerns the validity of assuming steady state conditions. It may be important to know the risk of high algal densities occurring for short periods. Short term variations in reactor behaviour can be caused, for example, by variations in inflow and load, intermittent recycling during periods of high wind and, possibly, variations in compartment model structure. The time scales associated with plankton growth rates compared to those for environmental fluctuations may be such that algal densities do reflect these changes. Grazing and other biological interactions, however, may dampen the algal response to fluctuating environmental conditions.

Algal growth is not controlled by nutrient availability alone, light energy being equally important. The same high winds that make recycled phosphorus available may increase turbidity and light extinction so reducing the potential for photosynthesis. Turbulent motion itself, by causing algae to move up and down may create an oscillating light climate. Growth in open water habitats is, primarily, rate controlled, although it is possible to envisage events causing sudden falls in population. Any upper limit to populations is likely to stem from growth rates falling to zero than from any limitation in available habitat.

A map showing sediment types and on which different stability regimes may be identified is the best single indicator of the nature of benthic habitats. The advantage of such maps is that they provide information integrated over time and on which other information can be superimposed. This can include features of the hydraulic structure such as the limits of wave action on the bed and their related frequency of occurrence, the area of bed below the thermocline and the areas of upwelling and downwelling in the general circulation. Indicating the limits of light penetration and the consequences of level fluctuation can also help to clarify the nature of the lake bed environment. Rate, event and space control can all affect the growth of benthic populations. Their influence is analogous to that already described for rivers although it is no longer simply a matter of determining species distribution in one dimension.

References

Aalderink, R. H., Lijklema, L., Breukelman, J., van Raaphorst, W. & Brinkman, A. G. (1984). Quantification of wind induced resuspension in a shallow lake. *Water Science and Technology*, **17**, 903–14.

Bagnold, R. A. (1962). Auto-suspension of transported sediment. *Proc. Roy. Soc. (Ser. A)*, **265**, 315–19.

Bagnold, R. A. (1966). An approach to the sediment transport problem from general physics. *United States Geological Survey Professional Paper*, **422I**, 1–37.

Banks, R. B. (1974). A mixing cell model for longitudinal dispersion in open channels. *Water Resources Research*, **10**, 357–8.

Banks, R. B. (1975). Some features of wind action in shallow lakes. *Proc. Am. Soc. Civ. Engrs*, **101** (EE5), 813–27.

Bathurst, J. C. (1978). Flow resistance of large scale roughness. *Proc. Am. Soc. Civ. Engrs*, **HY12**, 1587–603.

Bengtsson, L. (1973). Conclusions about turbulent exchange coefficients from model studies. Hydrology of lakes symposium, Helsinki. International Association of Hydrological Sciences, Publication No. 109, pp. 306–12.

Blanton, J. O. (1973). Vertical entrainment into the epilimnia of stratified lakes. *Limnol. & Oceanogr.*, **18**, 451–8.

Bowden, K. F. & Fairbairn, L. A. (1956). Measurements of turbulent fluctuations and Reynolds stresses in a tidal current. *Proc. R. Soc. (Ser. A)*, **214**, 422–38.

Bridge, J. S. (1981). Hydraulic interpretation of grain size distribution using a physical model for bed load transport. *Journal of Sediment Petrology*, **51**, 1109–24.

Bruce, J. P. & Clark, R. H. (1966). *Introduction to Hydrometeorology*. Pergamon Press, Oxford.

Bunker, A. F. (1976). Computation of surface energy flux and annual air–sea interaction cycles of the North Atlantic Ocean. *Monthly Weather Review*, **104**, 1122–40.

Buranathanitt, T., Cockrell, D. J. & John, P. H. (1982). Some effects of Langmuir circulation on the quality of water resource systems. *Ecological Modelling*, **15**, 49–74.

Bye, J. A. T. (1965). Wind driven circulation in unstratified lakes. *Limnol. & Oceanogr.*, **10**, 451–8.

Calvert, S. E. (1972). Sediment survey of Loch Leven, Kinross. Grant Institute of Geology, University of Edinburgh. Mimeograph.

Carling, P. A. (1983). Threshold of coarse sediment transport in broad and narrow natural streams. *Earth Surface Processes and Landforms*, **8**, 1–18.

Carper, G. L. & Bachmann, R. W. (1984). Wind resuspension of sediments in a prairie lake. *Canadian Journal of Fisheries and Aquatic Sciences*, **41**, 1763–7.

CERC (1977). *Shore Protection Manual, Vol. 1.* US Army Coastal Engineering Research Centre. Fort Belvoir, Virginia.

Chambers (1974). *Dictionary of Science and Technology*, Revised Edition. Edinburgh.

Chandler, T. J. & Gregory, S. (1976). *The Climate of the British Isles*. Longman, London.

Charlton, F. G. (1975). An appraisal of available data on gravel rivers. Report INT 151, Hydraulics Research Station, Wallingford.

Colebrook, C. F. & White, C. M. (1937). Experiments with fluid friction in roughened pipes. *Proc. R. Soc. (Ser. A)*, **161**, 367–81.

Cowan, W. L. (1956). Estimating hydraulic roughness coefficients. *Agricultural Engineering*, **37**, 473–5.

Darbyshire, J. & Darbyshire, M. (1957). Seiches in Louch Neagh. *Quarterly Journal of the Royal Meteorological Society*, **83**, 93–102.

Deacon, E. L. & Webb, K. E. (1962). Interchange of properties between sea and air: Small scale interactions. In *The Sea, Vol. 1*, ed. M. N. Hill. Interscience Publishers, New York.

Defant, F. (1918). Neue Methode zur Ermittlung der Eigenschingungen (Seiches) von Abgeschlossenen Wassermassen. *Annaler Hydrographie*, **46**, 78–85.

Defant, F. (1961). *Physical Oceanography*. Pergamon Press, Oxford.

DiGiano, F. A., Lijklema, L. & Van Straten, G. (1978). Wind induced dispersion and algal growth in shallow lakes. *Ecological Modelling*, **4**, 237–52.

Dillon, P. J. & Rigler, F. H. (1975). A simple method of predicting the capacity of a lake for development based on lake trophic status. *Journal of the Fisheries Research Board of Canada*, **32**, 1519–31.

Edwards, A. M. C. (1973). The variation of dissolved constituents with discharge in some Norfolk rivers. *Journal of Hydrology*, **18**, 219–42.

Einstein, H. A. (1950). The bed load function for sediment transportation in open channel flows. Technical Bulletin No. 1026. US Department of Agriculture, Soil Conservation Service.

Ekman, V. W. (1905). On the influence of the earth's rotation on ocean currents. *Ark. Mat. Astr. Fys.*, **2**, 1–52.

Elliot, J. M. (1971). Some methods for the statistical analysis of samples of benthic invertebrates. Freshwater Biological Association, Scientific Publication No. 25.

Ewing, G. (1950). Slicks, surface films and internal waves. *Journal of Marine Research*, **9**, 161–7.

Farrell, G. J. & Stefan, H. G. (1988). Mathematical modeling of plunging reservoir flows. *Journal of Hydraulic Research*, **26**, 525–37.

Ferguson, R. I. (1981). Channel forms and channel changes. In *British Rivers*, ed. J. Lewin. George Allen and Unwin, London, pp. 90–125.

Ferguson, R. I. (1986). River loads underestimated by rating curves. *Water Resources Research*, **22**, 74–6.

Fischer, H. B., List, E. J., Koh, R. C. Y., Imberger, J. & Brooks, N. H. (1979). *Mixing in Inland and Coastal Waters*. Academic Press, New York.

Francis, J. R. D. (1958). *A Textbook of Fluid Dynamics for Engineering Students*. Edward Arnold, London.

George, D. G. (1981). Wind induced water movements in the South Basin of Windermere. *Freshwater Biology*, **11**, 37–60.

George, D. G. & Edwards, R. W. (1976). The effect of wind on the distribution of chlorophyll A and crustacean zooplankton in a shallow eutrophic reservoir. *Journal of Applied Ecology*, **13**, 667–90.

George, D. G. & Heaney, S. I. (1978). Factors influencing the spatial distribution of plankton in a small productive lake. *Journal of Ecology*, **66**, 133–55.

Gibson, C. E. (1987). Sinking rates of planktonic diatoms in an unstratified lake: a comparison of field and laboratory observations. *Freshwater Biology*. **14**, 631–8.

Gilbert, G. K. (1914). The transport of debris by running water. *US Geological Survey Professional Paper*, **86**, 1–363.

Gorham, E. (1958). The physical limnology of northern Britain: an epitome of the Bathymetrical Survey of the Scottish Freshwater Lochs, 1897–1909. *Limnol. & Oceanogr.*, **3**, 40–50.

Graf, W. H. (1971). *Hydraulics of Sediment Transport*. McGraw-Hill, New York.

Graf, W. H. & Acoroglu, E. R. (1966). Settling velocities of natural grains. *Bulletin of the International Association of Scientific Hydrology*, **11**, 27–43.

Hack, J. T. (1957). Studies of longitudinal stream profiles in Virginia and Maryland. *US Geological Survey Professional Paper*, **294B**, 1–97.

Haines, D. A. & Bryson, R. A. (1961). An empirical study of wind factor in Lake Mendota. *Limnol. & Oceanogr.*, **6**, 356–64.

Hakanson, L. (1982). Lake bottom dynamics and morphometry: the dynamic ratio. *Water Resources Research*, **18**, 1444–50.

Hansen, N-E. O. (1978). Mixing processes in lakes. *Nordic Hydrology*, **9**, 57–74.

Heaps, N. S. (1984). Vertical structure of current in homogeneous and stratified waters. In *Hydrodynamics of Lakes*, CISM Lectures No. 286, ed. K. Hutter. Springer Verlag, Vienna, pp. 153–207.

Hilton, J. (1985). A conceptual framework for predicting the occurrence of sediment focussing and sediment redistribution in small lakes. *Limnol. & Oceanogr.*, **30**, 1131–43.

Hilton, J. & Rigg, E. (1985). A Pascal program for the calculation of effective fetches as used in wave height and frequency predictions. *Computers and Geosciences*, **11**, 493–500.

Holdgate, M. W. (1979). *A Perspective of Environmental Pollution.* Cambridge University Press, Cambridge.

Holling, C. S. (1973). Resiliency and stability of ecological systems. *Annual Review of Ecological Systems,* **4,** 1–24.

Horn, W., Mortimer, C. H. & Schwab, D. J. (1986). Wind induced internal seiches in Lake Zurich observed and modeled. *Limnol. & Oceanogr.,* **31,** 1232–54.

Hrbacek, J. (1966). A morphometrical study of some backwaters and fish ponds in relation to the representative plankton samples, with an appendix by C. O. Junge on depth distribution for quartic surfaces and other configurations. *Hydrobiological Studies,* **1,** 222–65.

Hutchinson, G. E. (1957). *A Treatise on Limnology.* John Wiley, New York.

Hutter, K. (1984). Linear gravity waves, Kelvin waves and Poincaré waves, theoretical modelling and observations. In *Hydrodynamics of Lakes,* CSIM Lectures, ed. K. Hutter. Springer Verlag, Vienna, pp. 41–73.

Imberger, J. (1980). Selective withdrawal: a review. *Proceedings of the Second International Symposium on Stratified Flows.* International Association for Hydraulic Research, Trondheim, pp. 381–400.

Imberger, J. & Hamblin, P. F. (1982). Dynamics of lakes, reservoirs and cooling ponds. *Annual Review of Fluid Mechanics,* **14,** 153–87.

Imboden, D. M. & Lerman, A. (1978). Chemical models of lakes. In *Lakes: Chemistry, Geology, Physics,* ed. A. Lerman. Springer Verlag, New York, pp. 341–56.

Ippen, A. T. & Harleman, D. R. F. (1952). Steady state characteristics of subsurface flow. *Circular US Bureau of Standards,* 521, No. 12, pp. 79–93.

Jarvis, P. G. (1976). The interpretation of the variations in leaf water potential and stomatal conductance found in canopies in the field. *Phil. Trans. R. soc. Lond. (Ser. B),* **273,** 593–610.

King, C. A. M. (1972). *Beaches and Coasts.* Edward Arnold, London.

Kirchner, E. B. & Dillon, P. J. (1975). An empirical method of estimating the retention of phosphorus in lakes. *Water Resources Research,* **11,** 182–3.

Knighton, D. (1984). *Fluvial Forms and Processes.* Edward Arnold, London.

Kolmogoroff, A. N. (1941). The local structure of turbulence in incompressible viscous fluids for very large Reynolds Numbers. *Dokl. Akad. Nauk.,* **30,** 301.

Kresser, W. & Laszloffy, W. (1964). Hydrologie du Danube. *La Houile Blanche,* **2,** 133–78.

Kullenberg, G., Murthy, C. R. & Westerberg, H. (1973). Experimental studies of the diffusion characteristics in the thermocline and hypolimnion regions of Lake Ontario. *Proceedings of the 16th Conference on Great Lakes Research,* International Association for Great Lakes Research, Ann Arbor, pp. 779–90.

Lacey, G. (1929). Stable channels in alluvium. *Proceedings of the Institution of Civil Engineers,* **229,** 259–384.

Lacey, G. (1958). Flow in alluvial channels with sandy mobile beds. *Proceedings of the Institution of Civil Engineers,* **27,** 145.

Lam, D. C. L. & Jacquet, J-M. (1976). Computations of physical transport and regeneration of phosphorus in Lake Erie, fall, 1970. *Journal of the Fisheries Research Board of Canada,* **33,** 550–63.

Lam, D. C. L. & Simons, T. J. (1976). Numerical computations of advective and diffusive transports of chloride in Lake Erie. *Journal of the Fisheries Research Board of Canada*, **33**, 537–49.

Lamb, H. (1945). *Hydrodynamics*, 6th edn. Dover Publications, New York.

Langmuir, I. (1938). Surface motion of water induced by wind. *Science*, **87**, 119–23.

Lathbury, A., Bryson, R. & Lettau, B. (1960). Some observations of currents in the hypolimnion of Lake Mendota. *Limnol. & Oceanogr.*, **5**, 409–13.

LaZerte, B. D. (1980). The dominating higher order vertical modes of the internal seiche in a small lake. *Limnol. & Oceanogr.*, **25**, 846–54.

Leenen, J. D. (1982). Wind induced diffusion in a shallow lake. *Hydrological Bulletin*, **16**, 231–40.

Leliavski, S. (1955). *An Introduction to Fluvial Hydraulics*. Constable, London.

Lemmin, H. & Imboden, D. M. (1987). Dynamics of bottom currents in a small lake. *Limnol. & Oceanogr.*, **32**, 62–75.

Lemmin, H. & Mortimer, C. H. (1986). Tests of an extension to internal seiches of Defant's procedure for determination of surface seiche characteristics in real lakes. *Limnol. & Oceanogr.*, **31**, 1207–31.

Leopold, L. B. & Maddock, T. (1953). The hydraulic geometry of stream channels and some physiographic implications. *US Geological Survey Professional Paper*, **252**, 1–57.

Leopold, L. B. & Wolman, M. C. (1957). Open channel patterns: braided, meandering and straight. *US Geological Survey Professional Paper*, **282B**, 39–85.

Leopold, L. B., Wolman, M. G. & Miller, J. P. (1964). *Fluvial Processes in Geomorphology*. W. H. Freeman, San Francisco.

Lerman, A. (1979). *Geochemical Processes*. Wiley Interscience, New York.

Livingstone, D. A. (1954). On the orientation of lake basins. *American Journal of Science*, **252**, 547–52.

Lord, S. R. (1966). Frequency curves and extension of records. *Journal of the Institution of Water Engineers*, **20**, 239–43.

Lorenzen, M. W. (1974). Predicting the effects of lake diversion on lake recovery. In *Modeling the Eutrophication Process*, ed. E. J. Middlebrooks, D. H. Falkenborg & T. E. Maloney. Ann Arbor Science, Ann Arbor, MI, pp. 205–10.

Maitland, P. S. & Smith, I. R. (1987). The River Tay: ecological changes from source to mouth. *Proceedings of the Royal Society of Edinburgh*, **92B**, 373–92.

May, R. M. (1981). Models for single populations. In *Theoretical Ecology: Principles and Applications*, 2nd edn, ed. R. M. May. Blackwell, Oxford, pp. 5–29.

McManus, J. & Duck, R. W. (1983). Side scan sonar recognition of subaqueous landforms in Loch Earn, Scotland. *Nature*, **303**, 161–2.

McNaughton, S. J. & Wolf, L. L. (1973). *General Ecology*, Holt, Rinehart and Winston, New York.

McNown, J. S. & Malaika, J. (1950). Effects of particle shape on settling velocity at low Reynolds numbers. *Transactions of the American Geophysical Union*, **31**, 74–82.

Metha, A. J. & Partheniades, E. (1975). An investigation of the depositional properties of flocculated fine sediments. *Journal of Hydraulic Research,* **13,** 361–82.

Miles, J. W. (1963). On the stability of heterogeneous shear flows. *Journal of Fluid Mechanics,* **16,** 209.

Minshall, G. W., Cumkins, K. W., Petersen, R. C., Cushing, C. E., Bruns, D. A., Sedell, J. R. & Vannote, R. L. (1985). Developments in stream ecosystem theory. *Canadian Journal of Fisheries and Aquatic Science,* **42,** 1045–55.

Monismith, S. G. (1985). Wind forced motions in stratified lakes and their effect on mixed layer shear. *Limnol. & Oceanogr.,* **30,** 771–83.

Morris, H. M. (1955). A new concept of flow in rough conduits. *Transactions of the American Society of Civil Engineers,* **120,** 373–98.

Mortimer, C. H. (1953). The resonant response of stratified lakes to wind. *Schweiz. Z. Hydrol.,* **15,** 94–151.

Mortimer, C. H. (1961). Motion in thermoclines. *Verh. Internat. Verein. Limnol.,* **14,** 79–83.

Mortimer, C. H. (1963). Frontiers in physical limnology with particular reference to long waves in rotating basins. Publication No. 10, Great Lakes Research Division, University of Michigan, pp. 9–42.

Murdie, G. (1976). Population models. In *Mathematical Modelling,* eds J. G. Andrews & R. R. McLane, Butterworths, London, pp. 98–105.

Murray, J. & Pullar, L. (1910). *Bathymetrical Survey of the Scottish Freshwater Lochs.* Challenger Office, Edinburgh.

Myer, G. E. (1969). A field study of Langmuir circulations. *Proceedings of the 12th International Conference on Great Lakes Research,* International Association for Great Lakes Research, Ann Arbor, pp. 652–63.

NERC (1975). Flood Studies Report, Vol. 1, Hydrological Studies. Institute of Hydrology, Wallingford, Natural Environment Research Council.

NERC (1980). Low Flow Studies, Report No. 1. Institute of Hydrology, Wallingford, Natural Environment Research Council.

Nikuradse, J. (1933). Stromungsgesetze in rauhen Rohren. *Verein Deutscher Ingenieure,* Forschungschaft No. 361.

Nixon, M. (1959). A study of the bankfull discharges of rivers in England and Wales. *Proceedings of the Institution of Civil Engineers,* **12,** 157–74.

Norman, J. D. (1964). Lake Vattern: investigations on shore and bottom morphology. *Geografiska Annaler,* **1–2,** 1–238.

Okubo, A. (1971). Ocean diffusion diagrams. *Deep Sea Research,* **18,** 789–802.

O'Loughlin, E. M. & Bowner, K. H. (1975). Dilution and decay of aquatic herbicides in flowing channels. *Journal of Hydrology,* **26,** 217–35.

Parde, M. (1947). *Fleuves et Rivieres.* Armand Collin, Paris.

Park, C. C. (1977). World wide variations in hydraulic geometry exponents: an analysis and some observations. *Journal of Hydrology,* **33,** 133–46.

Patterson, J. C., Hamblin, P. F. & Imberger, J. (1984). Classification and dynamic simulation of the vertical density structure of lakes. *Limnol. & Oceanogr.,* **29,** 845–61.

Petersen, J. C. & Mohanty, P. K. (1960). Flume studies of flow in steep rough channels. *Proceedings of the American Society of Civil Engineers*, **HY9**, 55–76.

Petts, G. E., Foulger, T. R., Gilvear, D. J., Pratts, J. D. & Thoms, M. C. (1985). Water movement and water quality variations during a controlled release from Kielder Reservoir, North Tyne River, UK. *Journal of Hydrology*, **80**, 371–89.

Pielou, E. C. (1969). *An Introduction to Mathematical Ecology*. Wiley Interscience, New York.

Platt, T. & Denman, K. L. (1975). A general equation for the mesoscale distribution of phytoplankton in the sea. *Memoires du Societe Royale des Sciences de Liege*, *6th ser.*, **7**, 31–42.

Prandtl, L. (1952). *The Essentials of Fluid Dynamics*. Blackie and Son, London, Glasgow.

Raudkivi, A. J. (1976). *Loose Boundary Hydraulics*, 2nd edn. Pergamon Press, Oxford.

Reynolds, C. S. (1988) Potamoplankton: paradigms, paradoxes and prognoses. In *Algae and the Aquatic Environment*, ed. F. E. Round. Biopress, Bristol.

Richards, T. L., Dragert, H. & McIntyre, D. R. (1966). Influence of atmospheric stability and over-water fetch on winds over the lower Great Lakes. *Monthly Weather Review*, **94**, 448–53.

Richardson, L. F. (1926). Atmospheric diffusion shown on a distance-neighbour graph. *Proc. R. Soc. (Ser. A)*, **110**, 709–27.

Rinaldi, S., Soncini-Sessna, R., Stehfest, H. & Tamura, H. (1979). *Modeling and Control of River Quality*. McGraw-Hill, New York.

Rzoska, J. (1978). *On the Nature of Rivers with Case Histories of Nile, Zaire and Amazon*. Dr W. Junk, The Hague.

Sangregorio, J. H. (1981). Evaluation of the mean residence time of water in monomictic lakes. European Applied Research Reports. Environment and Natural Resources Section, 1(2), pp. 353–84.

Schnoor, J. L. & diToro, D. M. (1980). Differential phytoplankton sinking and growth rates: an eigenvalue analysis. *Ecological Modelling*, **9**, 233–45.

Schumm, S. A. (1960). The shape of alluvial channels in relation to sediment type. *US Geological Survey Professional Paper*, **352B**, 17–30.

Schwarzenbach, J. & Gill, K. F. (1978). *System Modelling and Control*. Edward Arnold, London.

Scott, J. T., Myer, G. E., Stewart, R. & Walther, E. C. (1969). On the mechanism of Langmuir circulations and their role in epilimnion mixing. *Limnol. & Oceanogr.*, **14**, 495–503.

Scraton, R. E. (1984). *Basic Numerical Methods*. Edward Arnold, London.

Shellard, H. C. (1968). *Tables of Surface Wind Speed and Direction over the United Kingdom*. HMSO, London.

Shiau, J. C. & Rumer, R. R. (1974). Decay of mass oscillations in rectangular basins. *Proceedings of the American Society of Civil Engineers*, **100**(HY1), 119–36.

Shields, A. (1936). Anwendung der Ahnlichkeits-Mechanik und der Turbulenzforschung auf die Geschiebebewegung. *Preussische Versuchsanstalt fur Wasserbau und Schiffbau*. Berlin.

Schulman, M. D. & Bryson, R. A. (1961). The vertical variation of wind driven currents in Lake Mendota. *Limnol. & Oceanogr.*, **6**, 347–55.

Smith, C. A. B. (1969). *Biomathematics, Vol. 2*, 4th edn. Charles Griffin, London.

Smith, I. R. (1974). The structure and physical environment of Loch Leven, Scotland. *Proceedings of the Royal Society of Edinburgh*, **B74**, 81–9.

Smith, I. R. (1975). *Turbulence in Lakes and Rivers*. Freshwater Biological Association, Scientific Publication No. 29.

Smith, I. R. (1979). Hydraulic conditions in isothermal lakes. *Freshwater Biology*, **9**, 119–45.

Smith, I. R. (1982). A simple theory of algal deposition. *Freshwater Biology*, **12**, 445–9.

Smith, I. R. (1987). Control of population growth by events: an initial account. *Ecological Modelling*, **35**, 175–88.

Smith, I. R. & Lyle, A. A. (1979). *Distribution of Fresh Waters in Great Britain*. Institute of Terrestrial Ecology, Natural Environment Research Council.

Smith, I. R. & Sinclair, I. J. (1972). Deep water waves in lakes. *Freshwater Biology*, **2**, 387–99.

Smith, I. R., Lyle, A. A. & Rosie, A. J. (1981). Comparative physical limnology. In *Scotland's Largest Lochs*, ed. P. S. Maitland. Dr W. Junk, The Hague.

Smith, J. M. (1970). *Chemical Engineering Kinetics*, 2nd edn. McGraw-Hill Kogakusha, Tokyo.

Smith, R. V. (1976). *Agriculture and Water Quality*. Technical Bulletin No. 32, Ministry of Agriculture, Fisheries and Food, London.

Southwood, T. R. E. (1981). Bionomic strategies and population parameters. In *Theoretical Ecology, Principles and Applications*, 2nd edn, ed. R. M. May. Blackwell, Oxford, pp. 30–52.

Spiegel, R. H. & Imberger, J. (1980). The classification of mixed layer dynamics in lakes of small to medium size. *Journal of Physical Oceanography*, **10**, 1104–21.

Statzner, B. & Higler, B. (1985). Questions and comments on the river continuum concept. *Canadian Journal of Fisheries and Aquatic Sciences*, **42**, 1038–44.

Strahler, A. N. (1952). Hypsometric (area–altitude) analysis of erosional topography. *Bulletin of the Geological Society of America*, **63**, 1117–42.

Strahler, A. N. (1954). Quantitative geomorphology of erosional landscapes. *Comptes Rendues, 19th International Geological Congress*, Algiers, Sec. 13(3), pp. 341–54.

Straskraba, M. (1980). The effect of physical variables on freshwater production. In *The Functioning of Freshwater Ecosystems*, ed. E. D. Le Cren & R. H. Lowe-McConnell. Cambridge University Press, Cambridge, pp. 13–84.

Talling, J. F. & Rzoska, J. (1967). The development of plankton in relation to hydrological regime in the Blue Nile. *Journal of Ecology*, **55**, 657–72.

Tansley, A. G. (1935). The use and misuse of vegetational terms and concepts. *Ecology*, **16**, 284–307.

Terwindt, J. H. J. (1977). Deposition, transportation and erosion of mud. In *Interactions between Sediments and Fresh Water*, ed. H. L. Golterman. Proceedings of an International Symposium, Amsterdam, Sept. 1976. Dr W. Junk, The Hague, pp. 19–24.

Thackston, E. L. & Schnelle, K. B. (1970). Predicting effects of dead zones on stream mixing. *Proceedings of the American Society of Civil Engineers, Sanitary Engineering Division*, SA2, 319–31.

Thorpe, S. A. (1971). Asymmetry of the internal seiche in Loch Ness. *Nature*, 231, 306–8.

Thorpe, S. A. (1972). The internal surge in Loch Ness. *Nature*, 237, 96–8.

Thorpe, S. A., Hall, A. J., Taylor, C. & Allen, J. (1977). Billows in Loch Ness. *Deep Sea Research*, 24, 371–9.

Tilton, L. W. & Taylor, J. K. (1937). The accurate representation of the refractivity and density of distilled water as a function of temperature. *Journal of Research of the National Bureau of Standards*, 18, 205–14.

Trewartha, G. T. (1965). *An Introduction to Climate*, 4th edn. McGraw-Hill, New York.

Trewartha, G. T. & Korn, L. H. (1980). *An Introduction to Climate*, 5th edn. McGraw-Hill International, New York.

Turc, L. (1954). Le bilan d'eau des sols. Relations entre les precipitations, l'evaporation et l'ecoulement. *Annales Agronomiques*, 5, 491–596.

Twenhofel, W. H. (1950). *Principles of Sedimentation*. McGraw-Hill, New York.

US Army (1962). *Waves on Inland Reservoirs*. Technical Memorandum No. 132, Beach Erosion Board, Corps of Engineers, Washington, DC.

Vannote, R. W., Minshall, G. W., Cummins, K. W., Sedell, J. R. & Cushing, C. E. (1980). The river continuum concept. *Canadian Journal of Fisheries and Aquatic Science*, 37, 130–7.

Viner, A. B. & Smith, I. R. (1973). Geographical, historical and physical aspects of Lake George. *Proc. Soc. Lond. (Ser. B)*, 184, 235–77.

Vollenweider, R. A. (1968). The scientific basis of lake and stream eutrophication with particular reference to phosphorus and nitrogen as eutrophication factors. Technical Report, OECD, Paris, DAS/DS1/68, p. 27.

Vollenweider, R. A. (1975). Input–output models. With special reference to the phosphorus loading concept in limnology. *Schweizerische Zeitschrift fur Hydrologie*, 37, 53–84.

Walling, D. E. & Foster, I. D. L. (1975). Variations in the natural chemical concentration of river water during flood flows, and the lag effect: some further comments. *Journal of Hydrology*, 26, 237–44.

Walling, D. E. & Webb, B. W. (1983). The dissolved load of rivers: a global review. Hamburg Symposium. Dissolved loads of rivers and surface quantity/quality relationships. International Association for Scientific Hydrology, Publication No. 41, pp. 3–20.

Ward, P. R. B. (1977). Diffusion in lake hypolimnia. *Proceedings of the 17th Congress*, Vol. 2. International Association for Hydraulics Research, Baden-Baden, pp. 103–10.

Weringa, J. (1986). Roughness-dependent geographical interpolation of surface wind speed averages. *Quarterly Journal of the Royal Meteorological Society*, **112**, 867–89.

White, K. E. (1974). The use of radioactive tracers to study mixing and residence time distributions in systems exhibiting three dimensional dispersion. First European Conference on mixing and centrifugal separation, Cambridge. British Hydromechanics Research Association, Cranfield. Paper No. A6, pp. 57–76.

Wilcock, A. A. (1968). Koppen after fifty years. *Annals of the Association of American Geographers*, 12–68.

Williams, G. P. (1978). Bankfull discharge of rivers. *Water Resources Research*, **14**, 1141–54.

Wolman, M. G. (1955). The natural channel of Brandywine Creek, Pennsylvania. *US Geological Survey Professional Paper*, **271**, 1–56.

Wolman, M. G. & Leopold, L. B. (1957). River flood plains: some observations on their formation. *US Geological Survey Professional Paper*, **282C**, 87–109.

Wolman, M. G. & Miller, J. P. (1960). Magnitude and frequency of forces in geomorphic processes. *Journal of Geology*, **68**, 54–74.

Yang, P. C. & Wallis, S. G. (1987). The aggregated dead zone model for dispersion in rivers. *Proceedings of the Conference on Water Quality Modelling in the Inland Natural Environment*, British Hydromechanics Research Association, Cranfield, pp. 421–33.

Appendix 1: The Properties of Water

A.1 VISCOSITY–TEMPERATURE RELATIONSHIP

Poiseuille's original equation for absolute viscosity, μ, modified to SI units ($N\,s\,m^{-2}$ rather than Poise) is

$$= 1{\cdot}78\ 10^{-3}/(1 + 0{\cdot}033\ 75\theta + 0{\cdot}000\ 22\theta^2)$$

The equivalent equation for kinematic viscosity, v ($m^2\,s^{-1}$) is

$$= 1{\cdot}78\ 10^{-6}/(1 + 0{\cdot}033\ 75\theta + 0{\cdot}000\ 22\theta^2)$$

Values of kinematic viscosity

°C	0	5	10	15	20	25	30	35
$v \times 10^{-5}$	1·78	1·52	1·31	1·14	1·01	0·90	0·81	0·73

A.2 DENSITY–TEMPERATURE RELATIONSHIP

The table of water density at different temperatures published by Hutchinson (1957) is commonly accepted as standard. Examination of these data shows that the usual polynomial form of equation is not satisfactory, different parameters being required over different ranges. The equation for distilled water, proposed by Tilton and Taylor (1937) and adjusted for SI units is

$$\rho = [1 - (\theta - 3{\cdot}986\,3)^2/508\,929{\cdot}2 \,.\, (\theta + 288{\cdot}941\,4)/(\theta + 68{\cdot}129\,63)] \times 10^3$$

Table A.1 shows the difference between the Hutchinson values, adjusted to SI units, and those calculated from the above equation.

Table A.1

Temperature	Hutchinson value (kg m^{-3})	Equation value (kg m^{-3})	Difference
0	999·867 9	999·867 6	+0·000 3
4	1 000·000 0	1 000·000 0	0
10	999·727 7	999·728 1	−0·000 4
20	998·232 3	998·233 6	−0·001 2
30	995·675 6	995·678 3	−0·002 7

Appendix 2: Notation

To retain conventional usage, some duplication occurs but this should not cause any confusion.

a	Parameter in a number of empirical equations
a_f	Amplitude of a surface seiche
A_c	Catchment area at a distance L from the source
A_m	Catchment area at mouth
A'	Part of the catchment area where the altitude is greater than H'
A_r	Relative area of a catchment $(=A'/A_c)$
A_b	Cross-sectional area of a river channel at bankfull
A_h	Cross-sectional area of a river channel at depth h
A_L	Surface area of a lake
A_*	Area of a submerged body on which a drag force acts
B	Population number or biomass
B_0	Population at start
B_*	Carrying capacity
c	Parameter in a number of empirical equations
c	Wave celerity
c_o	Wave celerity in the shore zone
$c(s)$	Laplace transform of the output
$c(t)$	Output from a linear system
C	Chezy coefficient
C	Concentration of matter in water
C_q	Concentration in inflowing stream
C_0	Concentration at $t = 0$
C_s	Steady state concentration
C_a	Sediment concentration at reference level
C_w	Sediment concentration at the water surface

273

C_h	Sediment concentration at level h
C_D	Total drag coefficient
C_{FD}	Friction drag coefficient
C_L	Lift coefficient
C_*	Coefficient relating flood flow to catchment area
C'	Constant of integration
d	Size of sediment particle
d_{84}	Particle size in a sediment mixture, 84 per cent of which is smaller than d_{84}
d_{50}	Median sediment particle size
d_*	Diameter of a spherical particle
D	Total water depth
D_*	Depth of frictional resistance
D'	Coefficient of molecular diffusion
e	Base of natural logarithms ($=2\cdot7183$)
E	Mean annual actual evaporation
E_v	Entrainment velocity
f	Magnitude of an impulse function
$\sqrt{8/f}$	D'Arcy–Weisbach coefficient
F	Effective fetch
F_m	Drag force on a submerged body
g	Acceleration due to gravity
g'	Reduced gravitational acceleration ($=g\Delta\rho/\rho$)
G_d	Duration of gales
$G(s)$	Transfer function in a linear system
h	Depth of water measured upwards from the bed
h_b	Maximum water depth at bankfull
h_m	Maximum water depth
$\bar{h}$	Mean water depth ($=A_h/W_*$)
h_r	Hydraulic mean depth ($=A_h/P'$)
h'	Staff gauge reading
H	Height of river bed above datum
H_p	Height of highest point within a catchment
H_s	Height of source
H'	Altitude limit in a catchment
H_r	Relative height in a catchment ($=(H'-H)/(H_p-H)$)
H	Height of surface waves
H_o	Wave height in the shore zone
H_s	Significant wave height
H_{so}	Significant wave height at the shore zone limit

H_t	Height of growing waves
i	Slope of a lake surface
i_*	Tilt of the thermocline surface
j	Bank slope in a trapezoidal channel
j_c	Rate of increase of concentration with time
J	Quantity of heat transferred across unit area in unit time
J_*	Parameter in Turc's equation
$J(T)$	Fraction of the throughflow that has residence time less than T
$J'(T)$	Variable in the frequency curve form of the residence time distribution
k	Von Karman's constant $(=0.40)$
k_s	Seiche decay coefficient
k_*	Velocity decay coefficient in a wind drift current
k'	Rate coefficient in a first order reaction
K	Coefficient of eddy diffusion
K_h	Horizontal diffusion coefficient in lakes
K_s	Coefficient of sediment diffusion
K_x	Diffusion coefficient in the direction of motion
K_y	Lateral diffusion coefficient
K_z	Vertical diffusion coefficient in lakes
l	Mixing length
L	Distance along river from source
L'	Characteristic length dimension
L_m	Distance from source to mouth
L_f	Lift force on a submerged body
L_L	Lake length
$\bar{L}_q$	Mean of the log transformed discharges
$\bar{L}_y$	Mean of the transformed variable related to discharge
m	Concavity parameter
m	Mass of tracer injected
m_d	Mass of material deposited up to time t
m_L	Mass of material deposited per unit length
m_o	Mass of material in suspension at start
m_s	Mass of material in suspension at time t
M	Mass transferred across unit area in unit time
$1/n$	Manning coefficient
N	Coefficient of eddy viscosity
N_u	Number of streams of order u
N_*	Brunt–Vaisala frequency
P	Probability

P	Mean annual rainfall
P_r	Flow power, i.e. rate at which kinetic energy is introduced to a lake by inflowing water
P_w	Power per unit area transferred from wind to water
P'	Wetted perimeter
q	Discharge rate per unit width of flow
Q	Flow or discharge rate
Q_b	Bankfull discharge
q_{min}	Minimum recorded flow
$\bar{Q}$	Mean annual flow
r	Growth rate
r_h	Range of horizontal motion at the bed in shore zone waves
r_o	Mean annual runoff
r_p	Height of roughness projection
r_q	Ratio of the calendar day flow to the peak flow
r_s	Retention coefficient
r_w	Orbital radius of particle motion beneath deep water waves
R_e	Reynolds Number
R_{ed}	Reynolds Number for particles
R_i	Richardson Number
R_i^*	Modified Richardson Number
R_o	Rossby Number
s	Variable in the Laplace transform (frequency) domain
S	Slope of river bed
S_{1085}	Measure of channel slope
S_q	Suspended sediment load
S_*	Measure of thermocline stability
t	Time (as independent variable)
t_a	Actual storm duration
t_d	Minimum wind duration for steady state waves to develop
T	(Variable) residence time of a particle passing through a water body
T	Period of a surface wave
T_s	Significant wave period
T_t	Period of growing surface waves
T'	Primary period of a surface seiche
T'_n	Period of the nth harmonic of a surface seiche
T'_*	Period of the primary internal seiche
T_c	Column clearance time
T_e	Equalisation time

T_i	Return period, i.e. the mean interval between random events
T_p	Time to peak of a flood
T_r	Theoretical retention time $(=V/Q)$
T_s	Stabilisation time, i.e. the time for the concentration to reach 95 per cent of its long term value
T'_e	Interval between events
T_*	Interval associated with the occurrence of an event
u	Stream order number
u	Velocity fluctuation in the direction of the mean motion
u_m	Maximum current speed associated with the primary surface seiche
u'_m	Maximum current speed associated with the primary internal seiche
$u(s)$	Laplace transform of the input to a linear system
$u(t)$	Input to a linear system
U	Velocity of flowing water
U_f	Friction velocity
U_s	Velocity of a wind drift current at the surface
U_*	Characteristic velocity
U'	Velocity of a sediment particle in a flow
v	Transverse velocity fluctuation
v_h	Velocity of horizontal motion at the bed in shore zone waves
v_i	Variability index
v_w	Orbital velocity of particle motion beneath deep water waves
V	Volume of a water body
V_s	Still water settling velocity of a particle
w	Vertical velocity fluctuation
W	Wind speed
W_{hm}	Annual hourly maximum wind speed
W_{bx}	Top water width at bankfull
W_x	Top water width at depth h
W_*	Wedderburn Number
z	Distance from water surface, measured downwards
z_e	Thermocline depth
z_h	Thickness of the hypolimnion
z_m	Depth of the wave mixed layer
z_o	Roughness length
z_p	Depth of plunge point
α	Fractional population loss due to an event
α	Angle between wave crest and shore in deep water

α_o	Angle between wave crest and shore within the shore zone
β	Bifurcation ratio
γ	Angle between channel bed and the horizontal
δ'	Thickness of the laminar sublayer
θ	Temperature
θ	Entrainment function
θ_c	Critical value of entrainment function at the onset of bed movement
θ_s	Critical value of the entrainment function for sediment suspension
λ	Length of surface waves
λ_s	Wavelength of a seiche
λ_o	Wavelength in the shore zone
λ_{so}	Significant wavelength at the shore zone limit
λ_*	Measure of roughness spacing
μ	Coefficient of absolute viscosity
μ	Mean of a statistical variable
ρ	Density of water
ρ_e	Mean density of water in the epilimnion
ρ_h	Mean density of water in the hypolimnion
ρ_s	Density of a solid particle
σ	Standard deviation of a statistical variable
σ_q	Standard deviation of the log transformed discharges
σ_y	Standard deviation of a variable related to discharge
τ	Stress
τ	Time constant of a linear system
τ_c	Critical stress at the onset of bed movement
τ_o	Bed stress in a boundary layer flow
τ_s	Wind stress on water surface
υ	Coefficient of dynamic viscosity
ϕ	Coefficient of form resistance
ϕ	Latitude
χ	Coefficient of eddy conductivity
χ	Habitat coefficient
Ω	Angular velocity of the earth's rotation

Index

Actual evaporation, 16, 22
Advection, 39
Advection-diffusion equation, 54–6, 83, 93, 139, 197
Aggregated dead zone model, 140
Air-water interface, 158
Aquatic systems modelling, 246
Artificial channels, 140
Artificial roughness, 122

Backflow systems, 93–5
Balance equation, 200
Bankfull conditions, 101, 104, 106
Bankfull discharge, 33, 103
Bankfull flows, 103
Bars, 111–12
Base Flow Index, 19
Bed deposits, 62
Bed movement, 66–7
Bernoulli's equation, 43
Biological sampling, 245
Biomass, rate of change, 217
Block releases, 210–11
Bottom boundary layer, 47
Boundary layer, inverted, 52
Boundary layer flow, 48–51, 113–22
Brunt-Vaisala frequency, 177

Calendar day flow, 35
Carrying capacity, 239
Cascade model, 91
Catchment area, 10, 11, 18
  coefficient of fixation, and, 109–10
  river length, and, 9

Catchment characteristics and weather, 19
Catchment form, 8–12
Catchment surface, 17
Channel bed, micro-topography, 111–13
Channel depth, 104
Channel dimensions, 33–5, 126–7, 209
  flow relations, and, 101–8
Channel form
  parabolic, 106–7
  trapezoidal, 106–7
Channel geometry, 209
Channel length, 8
Channel slope, 8
Channel width and discharge, 104, 105
Chezy equation, 118–20
Climate
  classification, 19–22
  streamflow, and, 16–38
Coefficient of absolute viscosity, 40, 44
Coefficient of eddy conductivity, 45
Coefficient of eddy diffusion, 45
Coefficient of eddy viscosity, 45, 165
Coefficient of fixation and catchment area, 109–10
Cohesive sediments, 62, 72–3
Column model, 18
Compartment models, 88–97
Compensation flow, 210
Convergent flow, 194
Coriolis force, 163–5, 193
Correction factor, 123

D'Arcy-Weisbach equation, 118–20
Decay coefficient, 173
Deep water waves, 184, 186–91
Density gradients, 197
Density-temperature relationship, 271
Depth of frictional resistance, 164
Diffusion, 198
   see also Advection-diffusion equation
Diffusion coefficient, 67, 195, 198
Diffusion theory, 200
Dimensionless hypsometric curves, 11
Discharge capacity, 213
Discharge frequency, 26–31
Discrete events, 230
Dispersion coefficients, 56
Dispersion in rivers, 139–41
Dissolved solids concentration, 124–6
Dominant discharge, 103
Drag force, 50, 66
Drainage area, 14
Dredging, 211
Drift current decay coefficient and wind speed, 162
Drought curves, 38
Drought flows, 36
Duration of gales, 151
Dynamic ratio, 146, 201

Earth's rotation, 165, 179
Ecosystems
   definition, 215
   diagrammatic view, 215
   essential features, 215
   models, 215–19
Eddy spectrum, 46
Eddy viscosity, 45, 47, 51
Edge effects, 183
Ekman layer, 47
Ekman spiral, 163
Energy conversion, 118
Energy dissipation, 45
Energy distribution, 220
Energy inputs, 192
Energy loss, 118, 120, 173
Entrainment function, 129

Epilimnion, 95–7, 166, 174, 183
Equalisation time, 56
Erosion, 212
Erosion resistance, 62
Eulerian representation, 41
Evaporation, 16–17, 19
Event control, 254
Event controlled growth, 229–37
Event frequency, 255
Events
   physical, 230
   species response to, 231–4
Exponential growth, 217
Extreme value analysis, 32, 35, 38, 230

Fisheries investigations, 223
Flocculation, 63
Flood banks, 211
Flood control, 206, 213
Flood Studies Report, 32, 35, 38
Flood time base, 38
Flood time scale, 36
Floods, 31–8, 103, 209
Flow cycle, seasonal variation, 25–6
Flow duration curves, 30, 35, 37, 38, 129, 131
Flow equations, 118–19, 123
Flow rate, 19
Flow regime, changes in, 209, 210
Flow regulation, 204–10
Flow reversals, 38
Flow variability, 29
Fluid dynamics, 1, 39–73
   basic principles, 42
Fluid motion
   basic features, 40–3
   characteristics, 40–3
   graphical representation, 41
Food chain, 223
Freshwater ecosystems, 215–41
Friction drag coefficient, 50
Friction velocity, 71, 118

Geometric relations, 107
Gomperz curve, 9

Grampian Mountains, 36
Gravity force, 118
Growth forms, 218
Growth rate, 217, 218

Habitat availability, 239–40
Heat flux, 45
Higher harmonic oscillations, 173
Hydraulic geometry, 104, 128
Hydraulic mean depth, 118
Hydraulic radius, 108
Hydraulic stress, 65
Hydraulic systems, dynamics of,
    74–97
Hydroclimate, 1–3
    substance of, 1
Hydroclimate investigations, 244–6
Hydrodynamic effects, 227
Hydrodynamic stress, 234–5
Hydroelectric power schemes, 206
Hydrographs, 31
Hydrostatic balance, 161–2, 178
Hypolimnion, 95–7, 174, 178, 183

Impulse function, 78
Incomplete mixing, 82–8
Inflow dynamics, 191–4
Inorganic sediments, 60
Instrument availability, 245
Internal reaction rate, 252
Internal seiches, 178–9, 209
Invertebrates, 67
Inverted boundary layer, 52

Jarvis's equation, 253

Kinematic viscosity, 40, 45, 70, 271
Kinetic energy, 50, 118, 162, 238
Kinetic theory of gases, 46
Koppen climate classification, 23, 26
K-strategists, 255

Lagrangian representation, 42

Lake articulation, 147
Lake Chad, 22
Lake dimensions, 142
Lake ecology, 259–60
Lake ecosystems, idealised, 222–4
Lake Erie, 201
Lake models, 256–60
Lake Ontario, 198
Lake size and form measures, 142–6
Lake Windermere, 174
Lakes
    basic features of hydroclimate,
        256–9
    basic theory of motion, 156–7
    hydraulic characteristics, 142–203
    hydraulic structure, 155–7, 202
    hydrodynamics, 146–8, 156–7
    inflow and outflow dynamics,
        191–4
    irregularly shaped, 167–9
    isothermal, 163–73
    morphology, 142–8, 222
    motion in, 163–83
    origin and morphology, 142
    standard measures, 142–6
    stratified, 173–83
    turbulence in, 195–8
    wind characteristics, 148–55
Laminar flow, 41
Land drainage, 213
Langmuir circulations, 159–61
Laplace transforms, 75, 77, 91
Large scale bars, 111–12
Linear systems theorys, 74–82
Loch Leven, 142, 146, 168, 258
Loch Lomond, 147
Loch Ness, 142, 146, 147, 178, 180,
    185
Loch Tay, 6, 14
Locks, 211
Log transformed discharges, 129
Longitudinal river sections, 6–8
Low flow analysis, 38
Low flow periods, 38

Main stream volume, 139
Malthusian growth, 217

Manning equation, 118–20, 123, 128
Mass balance, 200, 209
  mixed basin, of, 79–82
Mass balance equations, 77, 132
Mass curve, 206
Mean depth, 118
Mean flow, 129
Mean free path, 46
Mean motion, 39, 43
Mean residence time, 85
Mean retention time, 95
Mean stream length, 14
Mean velocity, 104, 118, 128, 130,
  131, 138
Mean water depth, 108
Mean wind speed, 150, 151, 155
Michaelis-Menten kinetics, 235, 253
Michaelis-Menten relationship, 224
Mixed basin, mass balance of, 79–82
Mixing in rivers, 139–41
Mixing length concept, 46
Mixing length theory, 113, 120, 162
Mixing processes, 82–3
Molecular motion, 45
Momentum, 42

Natural discrete events, 230
Navier-Stokes equations, 42
Net overall growth rate, 229
Newton's Laws of Motion, 42
Newton's second law of motion, 44
Non-cohesive solid deposits, 62

Organic matter, 60, 62, 67, 223, 226
Outflow concentration, 82, 83, 90
Outflow dynamics, 191–4

Parabolic channel, 123, 127
Parallel systems, 89–91
Particle size, 60, 128
  variation, 108–11
Particulate matter, 131
  deposition of, 200
  direct accumulation, 200
  dissolved and suspended, 124–6

Particulate matter—*contd.*
  dynamics of, 64–73
  hydrodynamic properties of, 58–64
  suspension, in, 62–4, 67–70, 124–6
  return to water column, 200
  static properties, 60–2
Peak flow, 35
Peclet Number, 228–9
Phosphorus recycling, 201
Photosynthesis, 220, 256
Physical events, 230
Physical losses, 228–9
Phytoplankton, 223, 228, 238
Plankton, 222–3, 228, 244
  distribution, 238
Plankton growth, spatial variation in,
  237–9
Plants, 222–3
Plug flow, 69–70, 82, 83, 85, 87, 93,
  252, 253
Poisson approximation, 231
Population loss, 234–5, 254
  due to sudden discharge of toxin,
  235–7
Potential evaporation, 16
Primary food supply, 220
Principle of continuity, 42
Principle of energy conservation, 43

Quasi-steady state ecology, 255
Quenouille's correction, 30, 129, 131
Quickflow, 18

Rainfall, 18–19, 31
Rainfall-runoff relations, 25
Ramp function, 78
Rate control, 253
Rate controlled growth, 217, 224–9,
  231
Ratio of calendar day flow to peak
  flow, 35
Ratio of width-depth against water
  level, 104
Raw materials, 225–6
Reaction efficiency, 87–8
Recurrence intervals, 103
Recycling, 225–6

Regulators, 226–8
Relative roughness, 120, 138
Relative velocity, 138
Reservoirs
 abstraction, 205
 characteristic morphology, 208
 design, 206–7
 direct supply, 205
 effects on river habitats, 209
 features of, 208–9
 flood control, 205
 linked, 208
 objectives, 206
 regulating, 205
 types of, 204–6
 water level fluctuations, 207–8
Residence time distribution, 83–7,
 88–9
Residence time distribution function,
 85
Resources, 224–6
Resuspension, 201
Retention coefficient, 82
Reynolds Number, 40, 43, 50, 66, 70
Richardson Number, 174, 177, 180
River Avon, 28
River banks, 211
River basins
 basic features, 1
 interests involved, 242–4
 modification, 204–14
 structure and dynamics, 4–38
River beds, 111–13
River characteristics at a particular
 section, 128
River Continuum Concept, 219, 220
River Danube, 6, 109
River Dee, 36
River description, 247–9
River ecology, 219, 253–6
River ecosystems
 idealised, 219–22
 theory, 219
River Ganges, 26
River habitats, modifications to,
 209–14
River habits, transport and storage,
 249–53

River length, 8
 catchment area, and, 9
River manager, 244
River mechanics, 113–31
River models, 246–56
River networks, 12–16
River Nile, 26
River stability, 132–6
River Tay, 5, 6, 14
River Tweed, 6
Rivers
 cascade, as, 251
 changes in structure, 209, 211–14
 classification based on Ferguson's
  scheme, 101
 classification based on plan form,
  100–1
 classification of morphology,
  98–101
 dispersion and mixing in, 139–41
 hydraulic characteristics, 98–141
 hydrodynamic features, 131–41
 longitudinal sections, 6–8
 micro-hydraulics, 136–8
 morphology, 98–113
 numerical example, 126–31
 relations between channel
  dimensions and flow, 101–8
Rolling eddy, 58
Rossby Number, 164
Roughness flow, 120
r-strategists, 255
Runoff generation, 16–19

St Serf's Island, 169
Scaling factor, 11
Seasonal stratification, 95–7
Sediment distribution, 111
Sediment focussing, 201
Sediment movement, 132, 191
Sediment rating curve, 226
Sediment regimes, 146
Sediment stability, 131, 132, 146
Sediment-water exchange, 199–201
 spatial variation in, 200–1
Sediment-water interface, 179
Sedimentation, 228

Sensitivity analysis, 245
Series systems, 91–3
Settling velocity, 60, 63, 64, 71, 72
Shallow water waves, 184
Shear stress on a surface, 49
Sheilds' diagram, 70–2, 100
Shield's function, 109, 132
Shore zone waves, 189–91
Slope measure, 35
Solar heating, 156
Space controlled growth, 237–40
Spatial limits and scales, 245
Spatial variation in plankton growth, 237–9
Species distribution, 254
Species response to events, 231–4
Species-response curves, 224, 229, 231
Spiralling, 222
Staff gauge reading, 108
Stage-discharge curves, 123–4, 127
Standard input functions, 77
Steady state circulation in stratified lakes, 183
Step function, 78
Stokes Law, 50–1
Storage theory, 206–8
Stratification, 95–7, 156, 173–83
Stratified motion regimes, 179–82
Stratified profile, 176–8
Stream drift, 226
Stream ecosystem theory, 220
Stream order number system, 12–13
Streamflow and climate, 16–38
Streamlines, 194
Streamwater, chemical compositon, 18
Surface seiches, 169–73
Surface waves, 183–91
Suspended particulate matter, 62–4, 67–70, 124–6
Systems theory, 245

Tay Basin, 14
Tayside Region, 109
Temperature gradient, 45
Temperature response curves, 227

Theoretical retention time, 82–7
Thermocline, 174, 183
Thermocline tilt, 178–9, 209
Throughflow rate constant, 252
Toxic discharges, 230, 235–7
Transfer functions, 75, 77, 91
Trap data, 201
Tributaries, 222
Turbulence, 39, 40, 156, 188, 212
    basic features, 43
    criterion for occurrence of, 43
    free, 56–8
    horizontal, 197–8
    lakes, in, 195–8
    lateral, 53
    longitudinal, 53
    mixing length model of, 46
    nature of, 43–58
    thermally-driven, 150
    vertical, 51–3, 195–7
    wind speeds, and, 155
Turbulent coefficient, 68
Turbulent structure, 51–4
Turc's equation, 22
Two-dimensional flow, 46
Two-dimensional steady circulation, 165–6

Vegetation, 16
Velocity, 47, 113–18
Velocity depth profile, 162
Velocity duration curve, 130–1
Velocity fluctuations, 43, 44, 47, 53, 64, 71
Velocity gradient, 40, 46–7
Velocity profiles, 49, 66, 165
Viscosity, 40
Viscosity-temperature relationship, 271

Water depth, measures of, 107–8
Water levels, 127
Water movement
    characteristics, 39
    influence of 1, 146
    nature of, 1

Water properties, 271–2
Water quality, 223
  low flow periods, 38
Water temperature-density relation,
  176
Wave action, 146
Wave celerity, 183, 186, 191
Wave characteristics, 183–4, 189
  wind speed, and, 187
Wave crests, 158
Wave growth, 187
Wave height, 186
Wave length, 187
Wave motion, 183
Wave period, 187
Wave refraction, 190–1
Wave theory, 186
Weather and catchment
  characteristics, 19
Wedderburn Number, 180, 182, 193,
  209
Weirs, 212

Wentworth Scale, 60
Wind characteristics, 148–55
  land station data, 150–1
Wind conditions over water, 151–5
Wind drift velocity, 162
Wind force, 57, 58, 67
Wind speed, 150, 245
  associated changes with, 158
  diurnal variation, 150–1
  drift current decay coefficient, and,
    162
  mean, 150, 151, 155
  turbulence, and, 155
  wave characteristics, and, 187
Wind speed-duration curve, 150
Wind stress, 245
  water surface, on a, 50
Wind-stress relation, 158
Wind-water interactions, 157–62
  basic processes, 157–8
  empirical observations, 158–9
Wolman's equation, 120

Printed in the United States
By Bookmasters